BY THOMAS HERTOG

On the Origin of Time

BY THOMAS HERTOG, BARBARA BAERT,
AND JAN VAN DER STOCK

Big Bang: Imagining the Universe (anthology)

ON THE ORIGIN OF TIME

STEPHEN HAWKING'S FINAL THEORY

ON THE
ORIGIN
OF
TIME

THOMAS HERTOG

BANTAM | NEW YORK

Copyright © 2023 by Thomas Hertog

Published in the United States by Bantam Books, an imprint of Random House, a division of Penguin Random House LLC, New York.

BANTAM BOOKS is a registered trademark and the B colophon is a trademark of Penguin Random House LLC.

Illustration credits are located on pages 271–272.

LIBRARY OF CONGRESS CATALOGING-IN-PUBLICATION DATA
Names: Hertog, Thomas, author.
Title: On the origin of time : Stephen Hawking's final theory / Thomas Hertog.
Description: First edition. | New York : Bantam Books, an imprint of Random House, a division of Penguin Random House LLC, [2023] | Includes bibliographical references and index.
Identifiers: LCCN 2022035059 (print) | LCCN 2022035060 (ebook) | ISBN 9780593128442 (hardback) | ISBN 9780593128459 (ebook) | ISBN 9780593722626 (international)
Subjects: LCSH: Hawking, Stephen, 1942–2018. | Cosmology. | Universe.
Classification: LCC QB981 .H475 2023 (print) | LCC QB981 (ebook) | DDC 523.1/2—dc23/eng20221021
LC record available at https://lccn.loc.gov/2022035059
LC ebook record available at https://lccn.loc.gov/2022035060

Printed in the United States of America on acid-free paper

randomhousebooks.com

2 4 6 8 9 7 5 3 1

FIRST EDITION

Title-page image: ESA/Hubble & NASA, A. Newman, M. Akhshik, K. Whitaker
Book design by Simon M. Sullivan

To Nathalie

La question de l'origine cache l'origine de la question.
The question of the origin hides the origin of the question.
—François Jacqmin

PREFACE

THE DOOR TO STEPHEN HAWKING'S OFFICE WAS OLIVE GREEN, AND, though it was right off of the bustling common room, Stephen liked it to be slightly open. I knocked and entered, feeling as though I'd been transported into a timeless world of contemplation.

I found Stephen sitting quietly behind his desk, facing the entrance, with his head, too heavy to hold straight, leaning against a headrest on his wheelchair. He slowly raised his eyes and greeted me with a welcoming smile, as if he had been expecting me all along. His nurse offered me a seat next to him and I glanced at the computer on his desk. A screen saver scrolled perpetually across the screen: *To boldly go where Star Trek fears to tread.*

It was mid-June of 1998, and we were deep in the labyrinth of DAMTP, Cambridge's renowned Department of Applied Mathematics and Theoretical Physics. DAMTP was housed in a creaking Victorian building on the Old Press site on the banks of the river Cam, and for nearly three decades, this had been Stephen's base camp, the nexus of his scientific endeavors. It was here that he, wheelchair bound and unable to lift even a finger, had passionately strived to bend the cosmos to his will.

Stephen's colleague Neil Turok had told me the master wanted to see me. It was Turok's animated course, part of DAMTP's famous advanced math degree, that had recently kindled my interest in cosmology. Stephen had got wind, it seemed, that my exam results were excellent and wanted to see if I'd make a good doctoral candidate under his wing.

Stephen's dusty old office stuffed with books and scientific papers felt cozy to me. It had high ceilings and a large window that, I would later find out, he kept open even on freezing cold winter days. On the wall next to the doorway was a picture of Marilyn Monroe; below it a framed and signed photograph of Hawking playing poker with Einstein and Newton on the holodeck of the *Enterprise*. Two blackboards filled with mathematical symbols occupied the wall to our right. One featured a recent calculation to do with Neil and Stephen's latest theory of the origin of the universe but the drawings and formulas on the second one appeared to date from the early 1980s. Could they be his last handwritten scrawls?

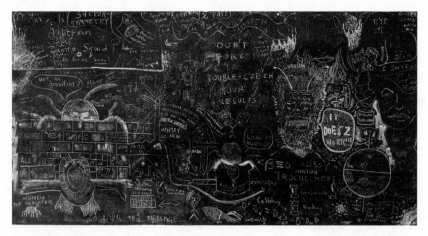

FIGURE 1. *This blackboard hung in Stephen Hawking's office at the University of Cambridge as a memento from a conference on supergravity he convened in June 1980. Filled with doodles, drawings, and equations, it is as much a work of art as a glimpse into the abstract universe of theoretical physicists. Hawking is drawn in the center near the bottom, with his back toward us.[1] (See color version, plate 10 in the insert.)*

A soft clicking broke the silence. Stephen had started talking. Having lost his natural voice in a tracheotomy following a bout of pneumonia more than a decade before, he now communicated through a disembodied computer voice. This was a slow, laborious process.

Mustering the last bit of force in his atrophied muscles, he exerted a feeble pressure on a clicking device, much like a computer mouse, which had been placed carefully in the palm of his right hand. The screen fitted to an arm of his wheelchair lit up, establishing a virtual lifeline between his mind and the outside world.

Stephen used a computer program called Equalizer that had a built-in database of words and a speech synthesizer. He appeared to navigate Equalizer's electronic dictionary instinctively, pressing the clicker rhythmically as if it were dancing to his brain waves. A menu on the screen displayed a number of frequently used words and the letters of the alphabet. The program's database included theoretical physics jargon, and the program anticipated his next word choice, displaying five options in the bottom row of the menu. Unfortunately, word selection was based on an elementary search algorithm, which failed to distinguish between general conversation and theoretical physics, with sometimes hilarious results, from cosmic microwave risotto to extra sex dimensions.

Andrei claims appeared on the screen below the menu. I waited, in hushed expectation, fervently hoping that I would understand whatever followed. A minute or two later Stephen directed the cursor to the icon "Speak" in the upper left corner of the screen and said, in his electronic voice, *Andrei claims there are infinitely many universes. This is outrageous.*

There we had it—Stephen's opening shot.

Andrei was the celebrated American-Russian cosmologist Andrei Linde, one of the founding fathers of the cosmological theory of inflation, proposed in the early 1980s. A refinement of the big bang theory, it postulates that the universe began with a brief burst of superfast expansion— inflation. Linde later concocted an extravagant extension of his theory, in which inflation produced not one but many universes.

I used to think of the universe as all there is. But how much is that? In Linde's scheme, what we have been calling "the universe" would be only a sliver of a vastly larger "multiverse." He envisaged the cosmos as an enormous swelling expanse of countless different universes lying far beyond one another's horizons, like islands in an ever-inflating ocean. Cosmologists were in for a wild ride. Stephen, the most adventurous of them all, had taken note.

Why worry about other universes? I asked.

Stephen's answer was enigmatic. *Because the universe we observe appears designed,* he said. Then, as he continued clicking, *Why is the universe the way it is? Why are we here?*

None of my physics teachers had ever spoken about physics and cosmology in such metaphysical terms.

"Isn't that a philosophical matter?" I tried.

"Philosophy is dead," Stephen said, eyes twinkling, ready to engage. I wasn't quite ready, but I couldn't help thinking that for someone who had renounced philosophy, Stephen used it liberally—and creatively—in his work.

There was a touch of magic about Stephen. With barely a flicker of motion, he breathed so much life into our conversation. He conveyed a magnetism and charisma that I had rarely seen. His broad smile and expressive face, simultaneously warm and playful, made even his robotic voice sound rich with personality and drew me deeper into the cosmic mysteries he pondered.

Like the Oracle of Delphi, he had mastered the art of packing a lot into a few words. The result was a unique way of thinking and talking about physics and, as I shall describe, a new physics altogether. But that concision also meant that even a minor clicking glitch such as a single missing word—"not," for instance—could, and often did, lead to frustration and confusion. That afternoon, however, I didn't mind being immersed in confusion, and I was thankful that Stephen's browsing of Equalizer gave me time to consider my responses.

I knew that when Stephen said that the universe appears designed, he was referring to the extraordinary observation that it emerged from its violent birth spectacularly well configured to sustain life—if billions of years in the future. This convenient fact has, in one way or another, bedeviled thinkers for centuries because it feels like a major fix. It's almost as if the geneses of life and the cosmos are entwined with each other, that the cosmos knew all along that one day it would be our home. What are we to make of this mysterious appearance of intent? It is one of the central questions humans ask about the universe and Stephen felt deeply that cosmological theory had something to say about it. The prospect—or hope—of being able to crack the riddle of cosmic design drove much of his work indeed.

This itself was exceptional. Most physicists prefer to steer away from such difficult, seemingly philosophical matters. Or they believe that one day it will turn out that the universe's delicately crafted architecture follows from an elegant mathematical principle at the core of the theory of everything. If this were the case, the universe's apparent design would seem like a lucky accident, a serendipitous consequence of objective and impersonal laws of nature.

But neither Stephen nor Andrei was your usual physicist. Reluctant to bet on the beauty of abstract mathematics, they felt that the uncanny fine-tuning of the universe that engendered life tapped into a deep problem at the roots of physics. Not content to merely apply the laws of nature, they sought a more expansive view of physics that included questioning the very origin of the laws. This led them to ponder the big bang, for it was presumably at the universe's birth that its law-like design was laid out. And it was on its birth that Stephen and Andrei strongly disagreed.

Andrei envisaged the cosmos as a gigantic ballooning space in which many big bangs continually produce new universes, each with its own physical properties, as if the latter were little more than our local cosmic weather. We should not be surprised to find ourselves in a rare universe suited to life, he argued, for we obviously couldn't exist in one of the many universes where life is impossible. Any impression of a grand design behind it all would be an illusion in Linde's multiverse, stemming from our limited view of the cosmos.

Stephen argued that Linde's grand cosmic extension, from universe to multiverse, was a metaphysical fantasy that didn't explain anything, although I sensed that he couldn't quite prove it. Nonetheless, I found it intriguing and exciting that the world's most eminent cosmologists, while strongly disagreeing, were debating these foundational questions with such strong conviction.

Doesn't Linde invoke the anthropic principle, the condition that we exist, to pick out a biofriendly universe in the multiverse? I ventured.

Stephen turned his eyes, looked at me, and slightly moved his mouth, leaving me puzzled. Later I would learn that this meant he disagreed. When he realized I hadn't been introduced to the sort of nonverbal layer of communication practiced within his inner circle, he turned his eyes back to the screen and set out to construct a whole new sentence. Two sentences, in fact.

The anthropic principle is a counsel of despair, he wrote, my bemusement mounting in sync with his clicking. *It is a negation of our hopes of understanding the underlying order of the universe, on the basis of science.*

Well, this was surprising. Having read *A Brief History of Time,* I was well aware that the early Hawking had frequently flirted with the anthropic principle as part of the explanation for the universe. A cosmologist at heart, Stephen had appreciated early on the surprising resonances between the large-scale physical properties of the universe and the existence of life as such. As far back as the early 1970s he had advanced an anthropic argument—wrongly, it turned out—as an explanation for why the expansion of the universe proceeded at the same rate in all three directions of space.[2] Had he changed his mind on the merits of anthropic reasoning in cosmology?

While Stephen took a medical pit stop to clear his trachea, I looked around his office. Copies of *A Brief History of Time* translated into exotic languages were piled high on a shelf that stretched across the length of the wall on our left. I wondered what else was in there that he no longer subscribed to. Next to these brief histories I noticed a row of his former graduate students' PhD dissertations. Starting in the early 1970s Stephen had established a celebrated school of thought at Cambridge, which had always included a small circle of rotating graduate students and postdoctoral scholars.

The titles of their dissertations touched on some of the most profound questions physics had grappled with in the late twentieth century. From the 1980s I saw Brian Whitt's *Gravity: A Quantum Theory?* and also Raymond Laflamme's *Time and Quantum Cosmology.* Fay Dowker's *Spacetime Wormholes and the Constants of Nature* took me to the early 1990s when Stephen and his colleagues thought wormholes—geometric bridges across space—influenced the properties of elementary particles. (Stephen's friend Kip Thorne would later put wormholes to use in the movie *Interstellar,* to get Cooper back to the solar system.) To Fay's right stood *Problems in M Theory* by Marika Taylor, Stephen's most recent academic offspring. Marika had worked under Stephen in the midst of the second string theory revolution when the theory morphed into a much larger web known as M-theory, and Stephen finally began to warm up to the idea.

All the way to the left on the shelf stood two copies of an older book with a thick green cover, *Properties of Expanding Universes*. This was Stephen's own PhD dissertation, going back to the mid-1960s, to the time when the large Holmdel Horn Antenna at Bell Telephone Labs picked up the first echoes from the hot big bang in the form of faint microwave radiation. Stephen proved in his thesis that if Einstein's theory of gravity was right, then the mere existence of these echoes meant time must have had a beginning. Now how did that square with Andrei's multiverse we were just talking about?

Immediately to the right of Stephen's, I saw Gary Gibbons's *Gravitational Radiation and Gravitational Collapse*. Gibbons was Stephen's first PhD student, in the early 1970s, during a time when the American physicist Joe Weber claimed to hear frequent bursts of gravitational waves coming from the center of the Milky Way. The intensity of gravitational radiation he reported was so high that it seemed the galaxy was losing mass at a rate that could not be sustained for eons—if this were true, there would soon be no galaxy left. Captivated by this paradox, Stephen and Gary toyed with the idea of constructing their own gravitational-wave detector in the basement of DAMTP. This was a narrow escape; rumors of gravitational waves turned out to be false and it would be another forty years before LIGO, the Laser Interferometer Gravitational-Wave Observatory, would finally succeed in detecting these elusive rippling vibrations.

Stephen usually took on one new graduate student every year to work with him on one of his high-risk high-gain projects, to do with either black holes—collapsed stars hidden behind a horizon—or with the big bang. He tried to alternate, assigning one student to work on black holes and the next one to work on the big bang so that at any time his circle of graduate students covered both strands of his research. He did this because black holes and the big bang were like yin and yang in his thinking—many of Stephen's key insights into the big bang can be traced to ideas he first developed in the context of black holes.

Both inside black holes and at the big bang, the macroworld of gravity truly merges with the microworld of atoms and particles. Under these extreme conditions, Einstein's relativity theory of gravity and quantum theory had better work together. Except they don't, and this is widely viewed as one of the biggest unsolved problems in physics. For example, both theories embody a radically different view of causality and determinism. Whereas Einstein's theory adheres to the old determinism of Newton and Laplace, quantum theory contains a fundamental element of uncertainty and randomness and retains only a reduced notion of determinism, about half of what Laplace thought it was. Over the years, Stephen's gravity group and its diaspora had done more than any research group in the world to expose the deep conceptual questions that arise when one tries to marry the seemingly contradictory principles of these two physical theories into a single harmonious framework.

Meanwhile Stephen was "sorted out," as his nurse put it, and had started clicking again. (A second pause in our conversation that afternoon involved watching a preview of an episode of *The Simpsons* in which Stephen appeared and that he had been asked to vet.)

I want you to work with me on a quantum theory of the big bang . . .

I had apparently arrived in a big bang year.

. . . *to sort out the multiverse.* He looked up at me with a broad smile, eyes twinkling again. This was it. Not by philosophizing or by an appeal to the anthropic principle but by weaving quantum theory deeper into cosmology were we to get a grip on the multiverse. The way he had put it made it sound like an ordinary homework problem, and though I could discern from his face that we had already started working, I had no clue in which direction spaceship Hawking was heading.

I am dying . . . appeared on the screen.

I froze. I glanced at his nurse who was reading quietly in a corner of the office. I looked back at Stephen, who seemed fine, as far as I could tell, and continued clicking away.

. . . *for* . . . *a* . . . *cup* . . . *of* . . . *tea.*

This was Britain and it was four P.M.

UNIVERSE OR MULTIVERSE? DESIGN(ER) OR NOT? This was the fateful question that would keep us occupied for twenty years. One homework problem led to another and soon Stephen and I found ourselves in the midst of what would become one of the most heated debates in theoretical physics in the first part of the twenty-first century. Nearly everyone had an opinion on the multiverse, though no one quite fathomed what to make of it. What started out as a doctoral project under his supervision evolved into a wonderfully intense collaboration ending only with Stephen's passing on March 14, 2018.

At stake in our work wasn't just the nature of the big bang, that enigma at the heart of existence, but also the deeper meaning of the laws of nature as such. What is it, ultimately, that cosmology finds out about the world? How do *we* fit into it? Such considerations take physics far out of its comfort zone. Yet this was exactly where Stephen liked to venture into and where his unmatched intuition, forged through decades of profound cosmological thinking, proved prophetic.

Like so many scholars before him, the early Hawking regarded the fundamental laws of physics as immutable, timeless truths. "If we do discover a complete theory . . . we would truly know the mind of God," he wrote in *A Brief History of Time*. More than ten years on, however, during our first meeting—and with Linde's multiverse breathing down our neck—I sensed he felt a crack in this position. Does physics really provide godlike foundations operating at the big bang origin of time? Do we need such foundations?

We were soon to discover that the Platonic pendulum in theoretical physics had swung too far indeed. When we trace the universe back to its earliest moments, we encounter a deeper level of evolution, at which the physical laws themselves change and evolve in a sort of meta-evolution. The rules of physics transmute in the primeval universe, in a process of random variation and selection akin to Darwinian evolution, with particle species, forces, and, we will argue, even time fading away into the big bang. Stronger still, Stephen and I came to see the big bang not only as the beginning of time but also as the origin of physical laws. At the heart of our cosmogony lies a new physical theory of the origin, which, we came to realize, at the same time encapsulates the origin of theory.

Working with Stephen was a voyage not only to the fringes of space and time but also deep into his mind—into what made Stephen Stephen. Our shared quest meant we grew close. He was a true seeker. Being around him, one could not fail to be influenced by his determination, and by his epistemic optimism that we could tackle these mystifying cosmic questions. Stephen made us feel like we were writing our own creation story, which, in a sense, we did.

And physics was fun! With Stephen you never quite knew when work ended and the party began. His insatiable passion to understand was matched only by his zest for life and his spirit for adventure. In April 2007, a few months after his sixty-fifth birthday, he took part in a zero-gravity flight aboard a specially equipped Boeing 727, which he saw as a prelude to a trip to space, all while his doctors panicked about him crossing the Channel on the Eurostar to come visit me in Belgium.

Meanwhile, with his natural voice permanently silenced and too weak now to move even a finger, he nevertheless became the biggest science communicator of our age. Inspired by a deep sense that we are part of a grand scheme that is written across the sky, waiting, as it were, for us to unravel, he shared his joy for discovery with a worldwide audience. Midway through our collaboration he wrote a book, *The Grand Design*, which reflects our confusion at the time. In it Stephen clings to the anthropic principle, the multiverse, and the idea of a final theory of everything, down to its rivalry with a God-created universe. But *The Grand Design* also contains the first traces of the new cosmological paradigm that would crystallize in our work a few years later. Shortly before his death, Stephen told me that it was time for a new book. This is that book. In the next few chapters I describe our journey back to and into the big bang, and how this journey ultimately led Hawking to discard the multiverse and replace it with a startling new perspective on the origin of time, profoundly Darwinian in spirit and nature and offering a radically revised understanding of the grand cosmic design.

We would often be joined in our endeavors by the American physicist Jim Hartle, Stephen's longtime collaborator with whom in the early 1980s he had pioneered the subject of quantum cosmology. Over the years the pair acquired a real knack for seeing the universe through a quantum lens. Even the language between them embodied their quantum thinking, as if they were wired differently. For example, by "the universe" cosmologists usually mean the stars and the galaxies and the vast space around us. When Jim or Stephen said "the universe," they meant the abstract quantum universe, awash in uncertainty, with all its possible histories living in some sort of superposition. But it was precisely their thoroughly quantum outlook that eventually made a genuine Darwinian revolution possible in cosmology. The later Hawking took quantum theory seriously—very seriously indeed—and decided to run with it, employing it to rethink the universe on the very largest scales. Quantum cosmology would be *the* field of research where Stephen remained at the forefront till the end of his life.

When a while into our collaboration he lost the remaining strength in his hand to press the clicker he used to converse, Stephen switched to an infrared sensor mounted on his glasses that he activated by slightly twitching his cheek. But eventually this too became difficult. Communication slowed, from a few words per minute to minutes per word, before basically grinding to a halt, even as demand for his voice skyrocketed.[3] Here was the world's most celebrated apostle of science, unable to talk. But Stephen wouldn't give up. With our intellectual connection deepened through years of close collaboration we moved increasingly beyond verbal communication. Bypassing Equalizer, sensors, and clickers, I would position myself in front of him, clearly in his field of vision, and probe his mind by firing questions. Stephen's eyes would light up brightly when my arguments resonated with his intuition. We would then build on this connection, navigating and exploiting the common language and mutual understanding we had forged over the years. It is out of these "conversations" that, slowly but steadily, Stephen's final theory of the universe was born.

There are critical junctures in science when metaphysical considerations come to the fore, whether we like it or not. At such forks in the road we learn something profound, not only about the workings of nature but also about the conditions that make our practice of science possible and worthy, and about the worldview our discoveries might nurture. Physics' quest to grasp what makes the universe just right for life has brought us to one such critical fork. For it is, at its core, a humanist question, much bigger than science. This is about *our* origins. Stephen's final theory of the universe contains the kernel of a uniquely powerful reflection on what it can mean to be human in this biofriendly cosmos, as stewards of planet Earth. For this reason alone it may ultimately prove to be his greatest scientific legacy.

CONTENTS

AUTHOR'S NOTE

My numerous conversations with Stephen over a span of twenty years are faithfully and truly woven into the narrative. Quotes from Stephen that have also appeared in published form are cited in the endnotes.

ON THE ORIGIN OF TIME

CHAPTER I

A PARADOX

Es könnte sich eine seltsame Analogie ergeben, daß das Okular auch des riesigsten Fernrohrs nicht größer sein darf, als unser Auge.

A curious correlation may emerge in that the eyepiece of even the biggest telescope cannot be larger than the human eye.

—LUDWIG WITTGENSTEIN, *Vermischte Bemerkungen*

THE LATE 1990S WERE THE CULMINATION OF A GOLDEN DECADE OF discovery in cosmology. Long regarded as a realm of unrestrained speculation, cosmology—the science that dares to study the origin, evolution, and fate of the universe as a whole—was finally coming of age. Scientists all over the world were buzzing with excitement about spectacular observations from sophisticated satellites and Earth-based instruments that were transforming our picture of the universe beyond recognition. It was as if the universe was speaking to us. This posed quite a reality check for theoreticians, who were told to rein in their speculation and flesh out the predictions of their models.

In cosmology we discover the past. Cosmologists are time travelers, and telescopes their time machines. When we look into deep space we look back into deep time, because the light from distant stars and galaxies has traveled millions or even billions of years to reach us. Already in 1927 the Belgian priest-astronomer Georges Lemaître predicted that space, when considered over such long periods of time, expands. But it wasn't until the 1990s that advanced telescope technology made it possible to trace the universe's history of expansion.

This history held some surprises. For example, in 1998 astronomers discovered that the stretching of space had begun to speed up around five billion years ago, even though all known forms of matter attract and should therefore slow down the expansion. Since then, physicists have wondered whether this weird cosmic acceleration is driven by Einstein's cosmological constant, an invisible ether-like dark energy that causes gravity to repel rather than to attract. One astronomer quipped that the universe looks like Los Angeles: one-third substance and two-thirds energy.

Obviously, if the universe is expanding now, it must have been more compressed in the past. If you run cosmic history backward—as a mathematical exercise, of course—you find that all matter would once have been very densely packed together and also very hot, since matter heats up and radiates when it is squeezed together. This primeval state is known as the *hot big bang*. Astronomical observations since the golden 1990s have pinned down the age of the universe—the time elapsed since the big bang—to 13.8 billion years, give or take 20 million.

CURIOUS TO LEARN more about the universe's birth, the European Space Agency (ESA) launched a satellite in May 2009 in a bid to complete the most detailed and ambitious scanning of the night sky ever undertaken. The target was an intriguing pattern of flickers in the heat radiation left over from the big bang. Having traveled through the expanding cosmos for 13.8 billion years, the heat from the universe's birth reaching us today is cold: 2.725 K, or about –270 degrees Celsius. Radiation at this temperature lies mainly in the microwave band of the electromagnetic spectrum, so the remnant heat is known as the *cosmic microwave background* radiation, or CMB radiation.

ESA's efforts to capture the ancient heat culminated in 2013 when a curious speckled image resembling a pointillist painting decorated the front pages of the world's newspapers. This image is reproduced in figure 2, which shows a projection of the entire sky, compiled in exquisite detail from millions of pixels representing the temperature of the relic CMB radiation in different directions in space. Such detailed observations of the CMB radiation provide a snapshot of what the universe was like a mere 380,000 years after the big bang, when it had cooled to a few

-300 δT [μK]   300

FIGURE 2. *A sky map of the afterglow of the hot big bang imaged by the European Space Agency's Planck satellite, named after quantum pioneer Max Planck. The speckles of different shades of gray represent slight temperature variations of the ancient cosmic microwave radiation as it reaches us from different directions in the sky. At first sight these fluctuations look random, but a close study has revealed that there are patterns interlinking different regions on the map. By studying these, cosmologists can reconstruct the universe's expansion history to model how galaxies formed and even predict its future.*

thousand degrees. This was cold enough to liberate the primeval radiation, which has traveled unhindered through the cosmos ever since.

The CMB sky map confirms that the relic big bang heat is nearly uniformly distributed throughout space, although not quite perfectly. The speckles in the image represent minuscule temperature variations indeed, tiny flickers of no more than a hundred-thousandth of a degree. These slight variations, however small, are crucially important, because they trace the seeds around which galaxies would eventually form. Had the hot big bang been perfectly uniform everywhere, there would be no galaxies today.

The ancient CMB snapshot marks our cosmological horizon: We cannot look back any farther. But we can glean something about processes operating in yet earlier epochs from cosmological theory. Just as paleontologists learn from stone fossils what life on Earth used to be like, cosmologists can, by deciphering the patterns encoded in these fossil flickers, stitch together what might have happened before the relic heat map was imprinted on the sky. This turns the CMB into a cosmological Rosetta

Stone that enables us to trace the universe's history even farther back, perhaps as far back as a fraction of a second after its birth.

And what we learn is intriguing. As we will see in chapter 4, the temperature variations of the CMB radiation indicate that the universe initially expanded fast, then slowed down, and, more recently (about five billion years ago), began accelerating again. Slowing down appears to be the exception rather than the rule on the scales of deep time and deep space. This is one of those seemingly fortuitous biofriendly properties of the universe, for only in a slowing universe does matter aggregate and cluster to form galaxies. If it hadn't been for the extended near-pause in expansion in our past, there would, again, be no galaxies and no stars, and thus no life.

In effect, the universe's expansion history was at the center of one of the very first moments in which the conditions for our existence slipped into modern cosmological thinking. This moment occurred in the early 1930s, when Lemaître made a remarkable sketch in one of his purple notebooks of what he called a "hesitating" universe, one with an expansion history much like the bumpy ride that would emerge from observations seventy years later* (see insert, plate 3). Lemaître embraced the idea of a long pause in the expansion by considering the universe's habitability. He knew that astronomical observations of nearby galaxies pointed to a high expansion rate in recent times. But when he ran the evolution of the universe backward in time at this same rate, he found that the galaxies must all have been on top of one another no more than a billion years ago. This was impossible, of course, for Earth and the sun are much older than that. To avoid an obvious conflict between the history of the universe and that of our solar system, he imagined an intermediate era of very slow expansion, to give stars, planets, and life time to develop.

In the decades since Lemaître's pioneering work, physicists have continued to stumble across many more such "happy coincidences." Make but a small change in almost any of its basic physical properties, from the behavior of atoms and molecules to the structure of the cosmos on the largest scales, and the universe's habitability would hang in the balance.

---

* Lemaître would often jot down scientific insights at one end of his notebooks and scribble spiritual reflections at the other, leaving a few blank pages in the middle as if to avoid unnecessarily mixing science and religion.

Take gravity, the force that sculpts and governs the large-scale universe. Gravity is extremely weak; it requires the mass of Earth just to keep our feet on the ground. But if gravity were stronger, stars would shine more brightly and hence die far younger, leaving no time for complex life to evolve on any of the orbiting planets warmed by their heat.

Or consider the tiny variations, one part in a hundred thousand, in the temperature of the relic big bang radiation. Were these differences slightly larger—say one part in ten thousand—the seeds of cosmic structures would have mostly grown into giant black holes instead of hospitable galaxies with abundant stars. Conversely, even smaller variations—one millionth or less—would produce no galaxies at all. The hot big bang got it just right. One way or another it set off the universe on a supremely biofriendly trajectory, the fruits of which would not become evident until several billion years later. Why?

Other examples of such happy cosmic coincidences abound. We live in a universe with three large dimensions of space. Is there anything special about three? There is. Adding just a single space dimension renders atoms and planetary orbits unstable. Earth would spiral into the sun instead of tracing out a stable orbit around it. Universes with five or more large space dimensions have even bigger problems. Worlds with only two space dimensions, on the other hand, may not provide enough room for complex systems to function, as figure 3 illustrates. Three dimensions of space seems just right for life.

Moreover, this uncanny fitness for life extends to the universe's chemical properties, which are determined by the properties of elementary particles and the forces acting between them. For example, neutrons are a tad heavier than protons. The neutron-to-proton mass ratio is 1.0014. Had it

FIGURE 3. *It appears difficult for life to form, let alone sustain itself in a universe with only two dimensions of space. Obvious mechanisms for hunting and eating don't work.*

been the other way around, all the protons in the universe would have decayed into neutrons shortly after the big bang. But without protons there would be no atomic nuclei and hence no atoms and no chemistry.

Another example of this is the production of carbon in stars. As far as we know, carbon is essential for life. But the universe was not born with it. Rather, carbon is formed in the nuclear fusion that takes place inside stars. In the 1950s, the British cosmologist Fred Hoyle pointed out that the efficient synthesis of carbon from helium in stars rests on a delicate balance between the strong nuclear force that binds the atomic nuclei, and the electromagnetic force. If the strong force were just a fraction stronger or weaker—as little as a few percent—then the nuclear binding energies would shift, compromising the fusion of carbon and hence the formation of carbon-based life. Hoyle felt this was so strange that he said the universe looked like a "put-up job," as if "a super-intellect has monkeyed with physics as well as with chemistry and biology."[1]

But the most dizzying life-engendering fine-tuning concerns dark energy. The value of the dark energy density that we've measured is extraordinarily small—a stunning 10^{-123} of what many physicists would consider a natural value. Yet this smallness is precisely what made the universe "hesitate" for about eight billion years before dark energy was able to muster sufficient strength to accelerate the expansion. Already in 1987, Steven Weinberg pointed out that if the dark energy density were somewhat larger—say 10^{-121} times its natural value—then its repulsive effect would have been stronger and kicked in earlier, once again closing the cosmic window of opportunity to form galaxies.[2]

In short, as Stephen emphasized in our first conversation, it appears as if the universe has somehow been designed to make life possible. The celebrated writer and theoretical physicist Paul Davies has spoken in this context of the universe's Goldilocks factor: "Like the porridge in the tale of Goldilocks and the three bears, the universe seems to be 'just right' for life, in many intriguing ways."[3] And while this doesn't quite mean that the cosmos should be teeming with life, the judicious tunings that render it habitable at all are by no means superficial qualities of the world. Instead they are inscribed deeply in the mathematical form of the laws of physics. The masses and properties of the array of particles, the forces governing their interactions, and even the overall composition of the universe—all of which seem tailor-made to support some form of life—reflect the specific character of the mathematical relationships that define what physicists call the laws of nature. So the riddle of design in cosmology is that the fundamental laws of physics appear to be specifically engineered to facilitate the emergence of life. It is as if there is a hidden plot at work that weaves together our existence with the basic rules on which the universe runs. This seems incredible. And it is! What is this plot?

Now, I should stress that this is a highly unusual puzzle for theoretical physicists. Typically, physicists use the laws of nature to describe one or another phenomenon or to predict the outcome of an experiment. They also attempt to generalize the existing laws in order to bring a wider range of natural phenomena into their fold. But these questions of design lead us down a quite different path. They prompt us to reflect upon the deeper nature of the laws, and on how we fit into their scheme. The thrill of modern cosmology is that it provides a scientific framework in which we can hope to elucidate this biggest of all mysteries. For cosmology is the one area of physics where we are an inherent part of the problem we're trying to solve.

HISTORICALLY, THE APPARENT design of the world has been taken as evidence that there is an underlying purpose to the workings of nature. This view goes back to Aristotle, perhaps the most influential philosopher who ever lived. Also a dedicated biologist, Aristotle observed that many of the processes operating in the living world seemed filled with intent. If living things that lack reason have an agenda, he thought, there must be a Final Cause directing the cosmos as a whole. Aristotle's teleological argument was persuasive, logical, comforting, and to some extent borne out empirically; the world around us exudes endless examples of final causes at work, from a bird collecting branches to build a nest to a dog digging a hole in the garden to unearth a bone. It isn't surprising, then, that Aristotle's teleological views held up, largely unchallenged, for almost two millennia.

But then, in the sixteenth century, somewhere on the outskirts of the Eurasian landmass, the work of a small circle of scholars sparked the modern scientific revolution. Copernicus, Descartes, Bacon, Galileo, and their contemporaries stressed that our senses can betray us. They embraced the Latin dictum *Ignoramus,* literally, "We do not know." This shift in perspective has had repercussions reaching far and wide. Some even consider it the single most influential transformation in the approximately two hundred thousand years that humans have dwelled on this planet. Moreover, its full significance has yet to unfold. The immediate upshot of the scientific revolution, at least in scholarly circles, was to discard Aristotle's deeply entrenched teleological worldview and to replace it with the idea that nature is governed by rational laws, operating here and now, that we can discover and understand. Indeed, the very essence of modern science is that, having admitted ignorance, we can acquire new knowledge, by experimenting and observing and by developing mathematical models that organize these observations in general theories or "laws."

Paradoxically, however, the scientific revolution deepened the riddle of the universe's biofriendliness. Prior to the scientific revolution, a unity of sorts underpinned man's conception of the world: Both the animate and the inanimate worlds were thought to be guided by an all-encompassing purpose, divine or otherwise. The world's design was viewed as a manifestation of a grand cosmic plan that, naturally, assigned a privileged role to man. The ancient world model put forward by the Alexandrian astronomer Ptolemy in his book *Almagest,* for example, was as much geocentric as it was anthropocentric.

But with the advent of the scientific revolution, the fundamental nature of life's relationship with the physical universe became fraught with confusion. Nearly five centuries on, our bafflement at the fact that the supposedly objective, impersonal, timeless laws of physics are nearly perfectly suited for life is a clear manifestation of this confusion. So, although modern science successfully abolished the old dichotomy between the heavens and the earth, it created a formidable new rift, between the living and nonliving worlds, leaving man's perception of his place in the grand cosmic scheme gnawingly uncertain.

As a matter of fact, we might get a better appreciation for how man's views on the ontology of nature's laws have come to be what they are by returning to the deeper roots of the idea that there are laws. The very first intimations of rules governing nature emerged in sixth-century B.C. Miletus, at the Ionian school of Thales, in what is today the western part of Turkey. Miletus, the richest of the Greek Ionian cities, was founded at a natural harbor near where the river Meander flows into the Aegean Sea. There the legendary

Thales, much like modern scientists, was willing to look beneath the surface appearances of the world in order to pursue knowledge at a deeper level.

Thales had a pupil, Anaximander, who created what the Greeks came to call περι φυσεως ιστορια, the "inquiry into nature," hence physics.

FIGURE 4. *Relief portraying the ancient Greek philosopher Anaximander from Miletus. Twenty-six centuries ago, Anaximander laid the first stones of the long and winding scientific road to rethink the world.*

Anaximander is also regarded as the father of cosmology, for he was the first to conceive of the earth as a planet, a giant rock floating freely in empty space. Beneath the earth wasn't more earth, without limit, nor gigantic columns, he reasoned, but the same sky we see above our heads. In this way he gave depth to the cosmos, transforming it from a closed box— with the heavens above and the earth below—into an open space. This conceptual shift allowed one to imagine celestial bodies passing beneath the earth, paving the way for Greek astronomy. Furthermore, Anaximander wrote a treatise titled *On Nature,* which is lost but is thought to have contained the following fragment:[4]

All things originate from one another,
and vanish into one another,
according to necessity;
For they give to each other justice and recompense for their injustice;
In conformity with the ordinance of time.

In these few lines Anaximander articulates the revolutionary idea that nature is neither arbitrary nor absurd but governed by some form of law. This is science's founding assumption: Beneath the surface of natural phenomena lies an abstract but coherent order.

Anaximander did not elaborate on what form the laws of nature might take, other than drawing an analogy with civic laws regulating human societies. But his most famous pupil, Pythagoras, proposed a mathematical basis for the world's order. The Pythagoreans assigned a mystical significance to numbers and attempted to construct the entire cosmos out of numbers. Their idea that the world could be described in mathematical terms was taken up and championed by Plato, who made it one of the pillars of his theory of truth. Plato likened the world of our experiences to a shadow of a far superior reality of perfect mathematical forms that exists quite separately from the one we perceive. The ancient Greeks would come to believe, therefore, that even though we can't readily touch or see the underlying order of the world, we can deduce it through logic and reason.

Impressive as their theories might be, however, the ancients' speculations about nature have little in common with modern physics, neither in substance nor in method or style. For one thing, the early Greeks reasoned almost entirely on aesthetic grounds and on the basis of prior assumptions, with little or no effort to test them. It just didn't occur to them. Consequently their conception of "physics" and of a law-like scheme of things bears no resemblance to a modern scientific theory. In his last book, *To Explain the World,* the late Steven Weinberg argued that, from a contemporary viewpoint, the early Greeks are better thought of not as physicists or scientists or even philosophers but as poets, since their methodology differed fundamentally from what would qualify as scholarly practice today. Of course, modern physicists too see the beauty in their theories, and most are sensitive to aesthetic considerations in their research, but such considerations are no alternative to the verification of theories through experiment and observation, which are, after all, the key innovations of the scientific revolution.

Nonetheless, Plato's vision to mathematize the world would prove to be immensely influential. When, twenty centuries later, the modern scientific revolution got under way, its main players were directly motivated by their faith in the Platonic program to seek a hidden order underpinning the physical world in terms of mathematical relationships. "The great book of nature," Galileo wrote, "can be read only by those who know the language in which it was written. And that language is mathematics."[5]

Isaac Newton, alchemist, mystic, a difficult character but one of the most powerful mathematicians who has ever lived, consolidated the mathematical approach to natural philosophy with his *Principia,* arguably the most important book in the history of science. Newton got a good start on it in lockdown, during the plague of 1665, when the University of Cambridge shut down. A fresh BA, Newton went back to his mother's house and its apple orchard in Lincolnshire. There he thought about calculus, about gravity and motion, and split light with a prism, showing that white light is made up of the colors of the rainbow. But it wasn't until much later, in April 1686, that Newton presented the *Philosophiae Naturalis Principia Mathematica* to the Royal Society for publication, complete with three laws of motion and the law of universal gravitation. The latter is perhaps the most famous law of nature, stating that the force of gravity between two bodies is proportional to the masses of the bodies and diminishes with distance as the square of their separation.

Newton's demonstration in the *Principia* that the same universal principles underpin the workings of the divine heavens and of the imperfect human world around us marked a conceptual and spiritual break with the past. It is sometimes said that Newton unified the heavens and the earth. His synthesis of the planetary motions in a handful of mathematical equations transformed all previous pictorial descriptions of the solar system and signaled the transition from the age of magic into what became modern physics. Newton's scheme provided the general paradigm that all subsequent physics has followed. Unlike the "physics" of the ancient Greeks, which we hardly recognize, contemporary physicists feel entirely at home with Newton's physics.

One much-celebrated success of Newton's laws was the discovery of the planet Neptune in 1846. Earlier astronomers had seen that the path of Uranus deviated slightly from the orbit predicted by Newton's law of gravity. The Frenchman Urbain Le Verrier, seeking to explain this stubborn discrepancy, made the bold suggestion that this was due to an unknown planet, much farther out still, whose gravitational pull slightly influenced the trajectory of Uranus. Using Newton's laws, Le Verrier was able to predict where in the sky this unknown planet should be in order to account for the wobble in Uranus's orbit—provided Newton's laws were correct. Indeed, astronomers soon found Neptune, within one degree of where Le Verrier had pointed them. This became one of the most remarkable moments of nineteenth-century science. It was said that Le Verrier had discovered a new planet "with the tip of his pen."[6]

Stunning successes like these over a span of several centuries seemed to confirm Newton's laws as universal definitive truths. Already in the eighteenth century, the French mathematician Joseph-Louis Lagrange remarked that Newton had been lucky to live in that unique time in human history in which it was possible to discover *the* laws of nature. Newton himself, in fact, seems to have done little to quench this emerging myth. Steeped in a tradition of mysticism, he saw the elegant mathematical form of his laws as a manifestation of the mind of God.

It is this mathematical formulation of nature's laws that embodies what physicists today mean when they use the word "theory." Physical theories derive their utility and predictive power from the fact that they describe the real world in terms of abstract mathematical equations that one can manipulate in order to predict what will happen without actually observing or performing the experiment. And it works! From the discovery of Neptune to the sensing of gravitational waves to the prediction of new elementary particles and antiparticles, time and again the mathematical groundings of the laws of physics have pointed to new and surprising natural phenomena that were later observed. Deeply impressed with this power to predict, the Nobel laureate Paul Dirac famously promoted the exploration of interesting and beautiful mathematics as the preferred way to practice physics. Mathematics "takes you by the hand," he said, "to discover new physical theories."[7] Today's string theorists have mostly adopted Dirac's dictum in their search for a final unified theory—at times succumbing to the ancients' temptation to take the mathematical beauty of their framework as a warrant of its truth. More than one of string theory's pioneers has lyrically expressed that the theory was too beautiful a mathematical structure to be irrelevant to nature.

At a deeper level, however, we still don't quite grasp why theoretical physics works so unreasonably well. Why does nature conform to a system of subtle mathematical relationships operating beneath its surface? What do these laws really mean? And why do they take the form they do?

Most physicists continue to follow Plato on this point. They tend to conceive of the laws of physics as eternal mathematical truths, living not just in our mind but operating in an abstract reality that transcends the physical world. The laws of gravity or quantum mechanics, for example, are usually thought of as approximations of a final theory that exists somewhere out there, in a realm yet to be uncovered. So while physical laws emerged in the modern scientific age first and foremost as tools to describe patterns found in nature, they have, ever since Newton identified their mathematical roots, taken on a life of their own, acquiring a sort of reality that supersedes the physical world. To the early-twentieth-century French polymath Henri Poincaré, the notion of unconditional Platonic laws was an indispensable presupposition to do science at all.

While Poincaré's vision is interesting and important, it is also puzzling. How exactly do such socially distant laws, up in their Platonic realm, come together to govern a physical universe, let alone one that is wonderfully biofriendly? Crucially, the discovery of the big bang means this isn't "just" a philosophical question. In fact, if the big bang is truly the origin of time, then it would seem that Poincaré had better be right, for if the physical laws are to determine how the universe started, one would have thought they must have some sort of existence beyond time. Interestingly, therefore, the big bang theory drags what might have seemed mere metaphysical considerations into the domain of physics and cosmology. The theory confronts us with some of our presuppositions on what physical laws are ultimately about.

At the end of the day, the idea that the laws of physics somehow transcend the natural world risks leaving the origin of their extraordinary fitness for life utterly mysterious. Physicists adhering to this scheme can only hope that a powerful mathematical principle at the kernel of the final theory will one day explain their biophilic character. The present-day Platonists' answer to the riddle of design is that it will turn out to be a matter of mathematical necessity: The universe is the way it is because nature has no choice. To the extent that this is an answer, it has a bit of the feel of Aristotle's Final Cause, disguised as modern theoretical physics. Furthermore, leaving aside the fact that a final theory of this kind remains a distant dream, even if such a powerful mathematical principle were ever found, it would hardly elucidate *why* the universe happens to be so remarkably bio-friendly. No Platonic truth of any kind indeed could quite bridge the gap between the nonliving and the living world that the dawn of modern science created. Instead, we would have to conclude that life and intelligence are just happy coincidences in a fundamentally impersonal, ideal mathematical reality, and there would be little more to understand.

THESE PLATONIC LEANINGS on matters of design in physics and cosmology, while not obviously wrong, differ radically from how, ever since Darwin, biologists came to see design in the living world.

Goal-directed processes and seemingly purposeful design are ubiquitous in the biological world. Indeed, these formed the basis for Aristotle's teleological views of nature in the first place. Living organisms are fantastically complex. Even a single living cell contains a diverse array of molecular components cooperating beautifully to fulfill its many tasks. In larger organisms, an enormous number of cells work together in a coordinated manner to construct elaborate, purposeful structures like eyes and brains. Before Charles Darwin, people were unable to understand how physical and chemical processes could possibly have created such a stunning functional complexity, and they invoked a Designer to explain it. The eighteenth-century English clergyman William Paley likened the wonders of the living world to the workings of a watch. Like a watch, Paley argued, the marks of design in the biological world are too strong to be ignored, and "Design must have a Designer."[8] But Darwin's paradigm-shattering theory of evolution decisively eliminated such teleological thinking from biology. Darwin's profound insight was that biological evolution is a natural process and that simple mechanisms—random variation and natural selection—can account for the apparent design in living organisms without the need to invoke a Designer.

On the Galápagos Islands, Darwin found a variety of finches that differed in the size and shape of their beaks. Ground finches had strong beaks, effective for cracking nuts and seeds, while tree finches had sharp, pointed beaks that were well adapted for extracting insects. These and other data on his journey suggested to Darwin that different varieties of finches were related and had evolved over time under influence of their specific ecological niches. In 1837, fresh from his voyage to the Galápagos on the HMS *Beagle,* Darwin made a simple sketch in one of his red notebooks of an irregularly branched tree. This sketch, of an ancestral tree, captured the sweep of his profound, burgeoning theory that all living things on Earth are related and descended from a single common ancestor—the trunk of the tree—through a gradual and incremental process of environmental selection operating on randomly mutating replicators (see insert, plate 4).

The core idea of Darwinism is that nature doesn't look ahead—it doesn't anticipate what may be needed for survival. Instead, any trends, such as the changing shapes of beaks or the progressive growth of the length of a giraffe's neck, follow from environmental selection pressures that act over long periods of time to amplify useful traits.

"There is grandeur in this view of life," Darwin would write more than twenty years later, "with its several powers, having been originally breathed into a few forms or into one; and that, whilst this planet has gone cycling on according to the fixed law of gravity, from so simple a beginning endless forms most beautiful and most wonderful have been, and are being, evolved."[9]

Darwinism turned Paley's argument upside down by demonstrating that the watch doesn't require a Swiss watchmaker. It gave a thoroughly evolutionary description of the living world in which its apparent design—including the laws it obeys—are understood as emergent properties of natural processes, not as the result of some supernatural act of creation.

DESPITE THEIR BEAUTY and grandeur, however, biological laws are often regarded a tad less fundamental than their counterparts in physics. For while emergent law-like patterns may be enduring, no one thinks of them as timeless truths. Moreover, determinism and predictability have played much less of a principled role in biology. Newton's laws of motion are deterministic; they allow physicists to predict where objects will be at any moment in the future, based on their locations and velocities today (or at any moment in the past). In Darwin's scheme, the randomness of mutations in living systems means that nearly nothing can be determined in advance—not even the laws that one day might emerge. This lack of determinism imbues biology with a retrospective element: One can understand biological evolution only by looking at it backward in time. Darwin's theory doesn't detail the actual evolutionary path from earliest life to today's diverse and complex biosphere. It doesn't predict the tree of life, since that wasn't—and couldn't have been—its purpose. Instead, Darwin's genius lay in his delineation of general organizational principles, while leaving the specific historic record to phylogenetics and paleontology. That is, Darwin's theory of evolution recognizes that life as we know it is the joint product of law-like regularities *and* a particular history. Its utility lies in the fact that it enables scientists to retrospectively construct the tree of life, starting from our observations of the biosphere today and the hypothesis of a common ancestry.

Darwin's finches provide a case in point. Had Darwin reasoned forward in time and attempted to predict the various finch species on the Galápagos Islands, starting from the chemical environment on the prebiotic Earth, he would have failed completely. The existence of finches, or of any other species roaming our planet, cannot be deduced solely on the basis of the laws of physics and chemistry, because every branching throughout biological evolution involves a game of chance. Some chance outcomes are favored by environmental circumstances and get frozen in, often with dramatic consequences downstream. Such frozen accidents help to determine the character of the subsequent evolution and may even take the form of new biological laws. The birth of the Mendelian laws of inheritance, for example, rests on the outcome of the collective branchings that led to sexually reproducing organisms in the first place.

In figure 5, I depict a modern version of the phylogenetic tree of life based on ribosomal RNA sequence analysis showing the three domains—bacteria, archaea, and eukaryota—and their common ancestor at the base of the tree. Everything on this tree, from its molecular basis to its branches of finch species, encapsulates the complex and convoluted history of billions of years of chemical and biological "experimentation," making biology a predominantly retrospective science. As the evolutionary biologist Stephen Jay Gould expressed it, "If we rewound the history of life and played the tape again, the species, body plan, and phenotypes that would evolve could be entirely different."[10]

The randomness inherent in biological evolution extends to other levels

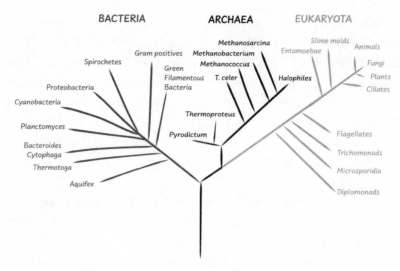

BACTERIA   ARCHAEA   EUKARYOTA

FIGURE 5. *Tree of life showing the three domains of life and, at the base of the tree, their last universal common ancestor—LUCA—the most recent life-form from which all life on Earth evolved.*

of history, from abiogenesis to human history. Like Darwin, historians account for accidental twists and turns in the course of history by distinguishing between describing "how" and explaining "why." To describe how, historians reason *ex post facto,* as biologists do, and reconstruct the series of specific events that led from one point to a given outcome. To explain why, however, requires one to think like a physicist and work forward in time in order to identify causal, deterministic connections that *predict* one particular historical path over any other. A superficial reading of history often appears to offer a causal deterministic explanation for why things happened one way and not another. But a more refined analysis tends to reveal an intricate web of competing forces at crossroads that, together with a great number of accidents, render the road taken far from obvious and certainly not inevitable, forcing one into a description of how, not why.

Consider the woods I can see from my office window, a few miles south of the battlefield of Waterloo. On June 17, 1815, on the eve of the main battle, Napoléon Bonaparte ordered his general, Emmanuel de Grouchy, to pursue the Prussian army in order to prevent them from merging with the Anglo-allied forces taking up position farther north. Grouchy dutifully marched northeast with a good part of the French troops but failed to find the Prussians. The next morning—from these woods I can see—he heard the deep sound of the French cannons in the distance and realized the battle had already begun. For a few crucial minutes he hesitated over whether to disobey Napoléon's orders by turning around and rushing to the aid of the remaining French troops. But he chose to walk on, away from destiny, in pursuit of the Prussians. Grouchy's decision at that moment is a remarkable frozen accident, one that not only impacted the battle's outcome but also affected the course of European history.

Or, to take another example, consider the rise of Christianity in the Roman Empire in the fourth century A.D. When the emperor Constantine ascended to the throne in A.D. 306, Christianity was little more than an obscure sect that was competing for influence with a range of cults. Why did Christianity take over the Roman Empire and rise to become the common faith? The historian Yuval Harari argues in his book *Sapiens* that there isn't a causal explanation and that the dominant influence of Christianity in Western Europe is better viewed as yet another frozen accident. Echoing Gould on biology, Harari asserts that "if we could rewind history and replay the fourth century a hundred times, we'd find that Christianity will have taken over the Roman empire a few times only." But the frozen accident of Christianity had far-reaching consequences. For one, monotheism encouraged a belief in God the creator, with a rational plan for the created world. It is hardly surprising, then, that when twelve centuries later modern science finally emerged in Christian Europe, the early scientists regarded their investigations as a sort of religious pursuit, setting the scene for the riddle of design we are still grappling with.

In general, the myriad of roads lying wide open at every point in history, from human history to biological and astrophysical evolution, mean that deterministic explanations are available at a rough, coarse-grained level only. At any stage of evolution, determinism and causality shape only the most general structural trends and properties, often on the basis of laws operating at a lower level of complexity. Human history, for example, rife with accidental twists and turns, has so far mostly played out within the confines of planet Earth, with the exception of a few excursions to other celestial bodies in the solar system. This planetary confinement is not surprising—and thus predictable—given the physical and geological environment in which human life arose. But it hardly tells us anything concrete about any epoch in human history.

Likewise, the order of the chemical elements and the structure of the periodic table of Mendeleev are essentially fixed by the laws of particle physics at the lower level. But the specific abundances of the elements on Earth are contingent on countless accidents leading up to and throughout its evolution.

Moving to the biochemical level, consider the rule that all life on Earth is DNA-based and that genes are made up of the four nucleotides abbreviated A, C, G, and T. The specific building blocks of the DNA molecule are probably the accidental outcome of abiogenesis on our planet. But the basic computational capacity that life must master in order to sustain itself runs deeper and may well determine the broad structural properties of the molecular carrier of genetic information on the basis of underlying mathematical and physical principles. This is borne out by the theoretical construction of self-reproducing automata by the Hungarian-American mathematician John von Neumann in 1948. Five years before Watson and Crick discovered the structure of DNA, von Neumann identified the critical computational problems that life must overcome to exist and posited an ingenious structure—seemingly the only possible structure—that achieves the capacity to replicate. The structure he drew is instantly recognizable as DNA.

Evolution continuously builds on an enormous chain of frozen accidents. Lower levels of complexity set the environment for higher levels of evolution, but this leaves so much room still for surprising turns that exceedingly improbable pathways are often realized, and determinism fails. The chance outcomes at countless branching events infuse a genuinely emergent element into evolution. They add a vast amount of structure and information that simply isn't contained in the lower-level laws, out of which new law-like patterns at higher levels can—and often do—emerge. For example, even though no serious scientist today believes there are special "vital forces" in biology that have no physicochemical origin whatsoever, the physical level alone does not determine what will be the laws of biology on Earth.

• • •

A mere eighteen days after the November 24, 1859, publication of his masterpiece, *On the Origin of Species,* Charles Darwin received a letter from the astronomer Sir John Frederick William Herschel. Herschel, the son of the discoverer of Uranus, expressed skepticism about the arbitrariness in Darwin's picture of evolution, saying his book was "the law of higgledy-piggledy."[11] Yet herein lay its strength. The beauty of Darwin's theory is that it offers a synthesis of the competing forces of random variation and environmental selection at work in the living world. Darwin found the sweet spot between the "why" and the "how" in biology, integrating causal explanations with inductive reasoning in a single coherent scheme. He showed that despite its fundamentally historical and accidental nature, biology can be a proper productive science that enhances our understanding of the living world.

Darwinism reinforced the scientific revolution and extended it into the one area where a teleological view had seemed unassailable—the living world. But the worldview it radiates couldn't be more different from the one emanating from fundamental physics. This difference shows up most clearly in their contrasting views of the riddle of design. Darwinism offers a thoroughly evolutionary understanding of the appearance of design in the living world. Physics and cosmology, on the other hand, have looked to the nature of timeless mathematical laws for an explanation of what made the transition to life possible in the first place. Scholars in the life sciences and physicists alike have often contrasted the "higgledy-piggledy" scheme of Darwinian evolution with the rigid and immutable character of the laws of physics. Not history and evolution but timeless mathematical beauty is thought to rule at the bottom level of physics. Lemaître's monumental insight that the universe expands obviously introduced strong evolutionary thinking into cosmology. Yet at a deeper level, where the fundamental origin of the apparent design is concerned, Lemaître's and Darwin's sketches in the color insert (plates 3 and 4, respectively) appear to exude profoundly different worldviews. It is the conceptual chasm that has separated biology and physics ever since the scientific revolution.

Bridging this rift had been on Stephen's mind since his earliest scientific endeavors but crystallized in a real research program only toward the end of the twentieth century, when much of his research efforts coalesced around the riddle of cosmic design. It was nothing less than his attempt, really, to change cosmology from within.

Let us return to these golden years. The unexpected observational discovery that the universe was accelerating tied in with equally surprising theoretical developments that suggested that the laws of physics might not be carved in stone after all. Evidence grew that at least some properties of the physical laws might not be mathematical necessities but accidental, reflecting the particular manner in which this universe had cooled from its hot big bang. From the species of particles to the strength of forces to the amount of dark energy, it became apparent that the universe's biofriendly bar code might not have been inscribed into its basic architecture right at the start, as if it were some sort of birth certificate, but was rather the result of an ancient evolution hidden deep in the hot big bang era.

Soon string theorists began to envisage a variegated multiverse, an enormous inflating space that would contain a patchwork of island universes, each with its own physics. This grand cosmic extension led to a sweeping change of perspective on matters of cosmic fine-tuning. Rather than mourning the demise of the dream of a unique final theory that predicts how the world should be, multiverse proponents attempted to turn that embarrassing failure around by transforming cosmology into an environmental science (albeit one with a very large environment!). One string theorist likened the local character of the physical laws in the multiverse to the U.S. East Coast weather: "tremendously variable, almost always awful, but lovely on rare occasions."[12]

We can get a feel of the magnitude of this change from the history of science. In the year 1597, the German astronomer Johannes Kepler came up with a model of the solar system based on the ancient Platonic solids, the five regular polyhedrons of which the cube is the best known. Kepler imagined attaching the approximately circular orbits of the six known planets to invisible spheres revolving around the sun. He then hypothesized that the relative sizes of these spheres are dictated by the condition that each sphere, other than the outermost one of Saturn, fits just inside one of the regular polyhedrons, and that each sphere, except for the innermost one of Mercury, fits just outside one of these.* Kepler's drawing in figure 6 illustrates this configuration. When Kepler placed the five geometric solids in the right order, one inside another, all fitting tightly together, he found that the nested spheres could be located at intervals corresponding to each planet's distance from the sun, together with Saturn moving along a sphere circumscribing the outermost polyhedron, with no room to change the relative radii. On these grounds he predicted the total number of planets—six—as well as the relative sizes of their orbits. For Kepler, the number of planets and their distances from the sun were a manifestation of a deep mathematical symmetry in nature. His *Mysterium Cosmographicum* is an attempt, really, to reconcile the ancient Platonic dream of the harmony of the spheres with the sixteenth century's insight that the planets move around the sun.

In Kepler's time the solar system was generally considered to be pretty much the whole universe. No one knew that the stars were suns with their own systems of planets. So it was perfectly natural to assume that the planetary orbits were a most fundamental matter. Today we know there is no great significance to the number of planets or their distances from the sun. We understand that the constellation of planets in the solar system isn't unique or even special but the accidental outcome of its formation history out of a swirling nebula of gas and dust surrounding the protosun. In the last three decades astronomers have observed thousands of planetary systems with widely different orbital configurations. Some stars have Jupiter-sized planets revolving around in a matter of days, others have three or even more habitable Earthlike planets, and yet other planetary systems have two stars, making for a chaotic pattern of days and nights and many other curious celestial phenomena.

If we do live in a multiverse, then the laws of physics in our universe would befall the same fate as the planetary orbits in the solar system. It would be futile to follow in Kepler's footsteps and search for a deeper ex-

---

* In particular, moving outward from the sun, Kepler placed the sphere of Mercury, an octahedron, the sphere of Venus, an icosahedron, the sphere of Earth, a dodecahedron, the sphere of Mars, a tetrahedron, the sphere of Jupiter, a cube, and, finally, the sphere of Saturn.

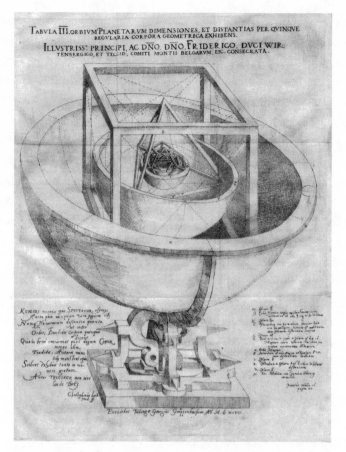

FIGURE 6. *In his first major astronomical work, the* Mysterium Cosmographicum, *Johannes Kepler put forward a Platonic model of the solar system relating the sizes of the (circular) orbits of the planets to the five regular solids. Four planetary spheres as well as the dodecahedron, tetrahedron, and cube are clearly visible in Kepler's drawing.*

planation of the fine-tunings that engender life. In the multiverse, the biofriendly properties of the laws would merely be the accidental result of random processes playing out in the hot big bang that gave rise to our particular island universe. Proponents of the multiverse argue that present-day Platonists have been looking in the wrong direction. It isn't a profound mathematical truth that renders the universe biofriendly, they hold, but simply excellent local cosmic weather. Any impression of a grand cosmic design is an illusion.

Yet there is a problem buried in this reasoning that will be of paramount importance when I come to discuss the core of Hawking's final theory: The multiverse is itself a Platonic construct. Multiverse cosmology postulates some sort of timeless metalaws governing the whole. But these metalaws do not specify in which one of the many universes *we* are supposed to be. This is a problem, for without a rule that relates the metalaws of the multiverse to the local laws within our island universe, the theory gets caught in a spiral of paradoxes that leaves one without verifiable predictions at all. Multiverse cosmology is fundamentally underdetermined and ambiguous. It lacks crucial *information* about our whereabouts in that crazy cosmic quilt and, as a consequence, it can't predict what we should see. The multiverse is like a debit card without a PIN or, worse, an IKEA closet without a manual. In a profound sense, the theory fails to say who we are in the cosmos, and why we are here.

Multiversists, however, don't give up easily. They have proposed a way to patch up the theory, a proposal so radical that it has shaken the scientific community ever since. This is the *anthropic principle.*

THE ANTHROPIC PRINCIPLE made its entry in cosmology in 1973. The astrophysicist Brandon Carter, who was a fellow student in Cambridge with Stephen, put forward the principle at a conference in Kraków in commemoration of Copernicus. This was a curious twist of history since in the sixteenth century, Copernicus had taken the first steps to demote mankind from its pivotal position in the cosmos.[13] More than four centuries later, Carter agreed with Copernicus that we humans are not central to the cosmic order. Yet, he reasoned, might we not be misled if we were to assume we are not special in any way, especially where our own observations of the cosmos are concerned? Perhaps we find the universe the way it is because we are here?

Carter had a point. Surely we wouldn't be observing anything where or when we didn't exist. Already in the 1930s scientists like Lemaître and the American astronomer Robert Dicke reflected on what properties the universe would need to have to support intelligent beings. Life-forms, intelligent or otherwise, rely on carbon, for example, which is fused by thermonuclear combustion in stars, a process that requires billions of years. But an expanding universe cannot provide billions of years of time unless it also contains billions of light-years of space. Therefore, we shouldn't be surprised, Lemaître and Dicke concluded, that we live in an old and big universe. Expanding universes have a preferred period in which astronomers made of carbon could be at work, and this necessarily influences what they could see.

Such conclusions are not fundamentally different from those we draw when we account for selection biases in everyday situations. But Carter went farther—much farther. He suggested that selection effects play out not just within one universe—ours—but across the multiverse. He suggested there is an anthropic principle at work, a rule above and beyond the impersonal metalaws governing the multiverse, that embodies the optimal cosmic conditions for life and that "acts" to select which one of the many universes should be ours.

That was a radical suggestion indeed. Once again maneuvering life in a privileged position at the center of the explanation of the universe, Carter's anthropic principle seems to take us back five centuries to the days before Copernicus. By positing a certain preferred state of affairs that includes life, intelligence, or even consciousness, it even flirts with teleology—the Aristotelian view that the scientific revolution had successfully overthrown, or so we thought.

Not surprisingly, then, when in 1973 Carter first put forward his cosmological anthropic principle, with theoretical evidence for any kind of multiverse patchy at best, his musings were generally dismissed as nonsense. But when toward the end of the century, in a remarkable turn of events, multiverse theory gained traction, Carter's anthropic thinking was swiftly resurrected and seized on to sort out our place in this vast cosmic patchwork. The anthropic principle came to be seen as the PIN that converted multiverse theory from an abstract Platonic edifice into a proper physical theory with real explanatory potential.

Multiverse aficionados declared that they had found a second possible answer to the mystery of the universe's design—the first being that it's just a coincidence, a lucky consequence of a deep but (as yet) mysterious mathematical principle at the core of existence. The new answer coming from anthropic multiverse cosmology was that the apparent design is a property of our "local" cosmic environment: We inhabit a rare biofriendly universe in a huge cosmic mosaic of island universes, singled out by the anthropic principle. Excitement at this development ran high. "We are together, the universe and us," Linde proclaimed. "I cannot imagine a consistent theory of the universe that ignores life and consciousness."[14] In his book *The Cosmic Landscape*, the opinionated string theorist Leonard Susskind of Stanford University (which can be relied upon for bold speculation) portrayed the tandem of objective metalaws together with a subjective anthropic principle as the new paradigm for fundamental physics.

The particle physics titan Steven Weinberg too suggested that anthropic reasoning signaled the dawn of a new era in cosmology. His unifying insight in the late 1960s, that the electromagnetic and weak nuclear forces are one and the same, had formed the basis of the Standard Model of particle physics. Some of the predictions of the Standard Model have since been verified to a stunning precision of no less than fourteen decimal places, making it the most accurately tested theory in physics ever. Yet for all this precision, Weinberg felt that to understand the deeper reasons why the Standard Model takes the particular form it does, we would need to supplement the mathematical principles of orthodox physics with a principle of an entirely different nature. "Most advances in the history of science have been marked by discoveries about nature," he told us in his "Living in the Multiverse" lecture at Cambridge, "but at certain turning points we have made discoveries about science itself and what we consider to be an acceptable theory. We may be at such a turning point . . . the multiverse legitimates anthropic reasoning as a new basis for physical theories."[15] The worldview Weinberg evokes here echoes a form of dualism. There are physical laws or metalaws, and we are discovering them, but they are cold and impersonal. In addition to these, however, there is the anthropic principle that, in its own mysterious way, bridges the (meta) laws with the physical world we experience.

THE BACKLASH HAS been fierce. Over the years the anthropic principle has become by any margin the most contentious issue in theoretical physics. Some are unequivocal in their opposition. "Inflationary theory dug its own grave," Princeton's co-discoverer of cosmological inflation, Paul Steinhardt, has declared. "It's like giving up," Nobel laureate David Gross of the University of California put it bluntly. Others feel that the entire discussion about our place in the cosmos is premature. "It's too soon to think about such matters,"[16] the otherwise visionary theorist Nima Arkani-Hamed told an audience of string theorists in the summer of 2019. Five centuries after the modern scientific revolution, which planted the seeds for dualism in physics, this is quite a remarkable statement.

To Stephen's frustration, a silent majority of theorists continued to look away—lost in math. Most theoretical physicists felt—and still feel—that an inquiry into the deeper origin of the universe's biofriendliness lies outside the scope of their discipline. They'd rather believe that the problem will somehow go away when we discover the master equation of string theory that rules the multiverse. One time over tea in DAMTP, Stephen, never shy about putting the cat among the pigeons, complained about this. "I am amazed," he said, "that people [string theorists] can have such blinkered vision, and not seriously ask how and why the universe got here."[17] Stephen held that it would not be enough, if we are to elucidate the mystery of design, to find the abstract mathematical metalaws. For him the search for a unified theory in physics was inextricably tied to our big bang origins. The dream of a final theory, he argued, cannot be achieved if we conceive of it as "just" another laboratory problem, but must be pursued in the context of cosmological evolution. Mathematics was Stephen's servant, not his master, in his pursuit of a new vision of the universe. So Hawking agreed with anthropic principlists that a better understanding of the universe's life-encouraging properties was important and that plain Platonism wouldn't suffice, that it would require a paradigm shift, a fundamental change in the way we conceive physics and the study of the universe.[18] However, he would grow increasingly skeptical that anthropic reasoning was the kind of revolutionary shift we needed in the light of these developments. His primary concern with the anthropic principle as part of a new cosmological paradigm wasn't so much its qualitative nature. Biology and other historical sciences are rife with predictions of a more qualitative kind. The real problem for him was that anthropic reasoning derails the basic scientific process of prediction and falsification.

This process was extensively discussed by the Austrian-British philoso-. pher of science Karl Popper. According to Popper, what makes science a uniquely powerful approach to acquiring knowledge is the fact that over and over again, consensus is reached among scientists as a result of ratio-

FIGURE 7(A). *In August 2001, Martin Rees, standing just to the left of Stephen, convened a meeting at his farmhouse in Cambridge, England, to debate the merits—if any—of the anthropic principle in fundamental physics and cosmology. It is in the margin of this conference that Stephen and the author (third row, behind Stephen) began to discuss in earnest how a quantum outlook on the cosmos might supplant anthropic cosmological reasoning. Rees's conference brought together many of our colleagues who'd play a key role along our journey, including Neil Turok (seated left), Lee Smolin (seated right) and Andrei Linde, who's standing all the way on the right in the middle row. Left of Linde we have Jim Hartle, barely visible behind Bernard Carr, and then Jaume Garriga, Alex Vilenkin, and Gary Gibbons.*

nal argument on the basis of available evidence. Popper realized that a scientific theory can never be proven true, but it can be falsified, meaning that it can be contradicted by experiments. But—and this is Popper's key point—this process of falsification has only been possible because theoretical hypotheses have been required to make unambiguous predictions such that, if contrary results were found, at least one premise of the theory would have been shown not to apply to nature. The reason this is central to the way science works is that this situation is asymmetric; confirmation of a theoretical prediction supports but doesn't prove a theory, while falsification of a prediction can show it to be false. The possibility of failure of a hypothesis always lurks around the corner in science, and that is an essential part of how it advances.

But the anthropic principle places this process on shaky ground, because one's personal criteria for what constitutes a biofriendly universe inject a subjective element into physics that compromises Popper's process of falsification. Your anthropic perspective might select one patch in the multiverse with this set of laws, whereas my anthropic leanings might pick out another patch, with a different set of laws, with no objective rule at hand to decide which one is correct.

This is much unlike Darwinian evolution, which ingeniously avoids anything like the would-be analogue of anthropic reasoning slipping into biology. Whether extraterrestrial life exists, let alone how it evolved, plays no role in Darwin's theory. Nor does Darwinism leave any room for singling out one particular species for a privileged role in biological affairs, be it *Panthera leo* or *Homo sapiens* or any other. Quite the contrary, Darwinism is rooted in our relationship with the rest of the living world. It recognizes the interconnectedness of it all. One of Darwin's monumental insights was that *Homo sapiens* co-evolved with everything else in the living world. "We must acknowledge, as it seems to me, that man with all his noble qualities . . . still bears in his bodily frame the indelible stamp of his lowly origin," he wrote in *The Descent of Man.* How profoundly different this is from Carter's anthropic principle in cosmology, operating outside the natural evolution of the universe, as if it were an add-on to it all.

In a Popperian sense, where it concerns falsification, the anthropic multiverse differs scarcely from the cosmology of the seventeenth-century German polymath Gottfried Leibniz. In his work *The Monadology,* Leibniz suggested there are infinitely many universes, each with its own space, time, and matter, and that we live in the best of all possible worlds, selected by God in all His goodness.

It is quite understandable then that the scientific community has found itself in perpetual disagreement on the merits of the anthropic principle. In his incisive critique of string theory, *The Trouble with Physics,* the American physicist and writer Lee Smolin pointedly notes that "once a non-falsifiable theory is preferred to falsifiable alternatives, the process of science stops and further increases in knowledge are ruled out." This was also what worried Stephen in our first conversation in his office, that once one buys into the anthropic principle, one gives up on the basic predictivity gained by science.

We have reached an impasse. The anthropic principle was meant to specify "who we are" in the vast cosmic patchwork and in this way function as a bridge connecting abstract multiverse theory to our experiences as observers within this universe. However, it fails to do so in a manner that upholds the basic principles of scientific practice, leaving multiverse cosmology with no explanatory power whatsoever.

This leads us to a remarkable observation: In the broadest sense we have made surprisingly little progress since the modern scientific revolution to fathom the deeper origin of the apparent design that underpins physical reality. Yes, we now understand the universe's expansion history in exquisite detail, we understand how gravity shapes the large-scale universe, and we understand the precise quantum behavior of matter down to scales much smaller than the size of a proton. But that detailed physical understanding, of enormous significance in its own right, has only served to accentuate the deeper-lying riddle of design. The universe's biophilic character continues to create confusion, dividing the scientific community and the broader public alike. A deep conceptual chasm continues to separate our understanding of the living world and that of the underlying physical conditions that make life possible at all. Why did the mathematical laws laid out at the big bang turn out to be fit for life? What should we make of the fact that they did? The rift separating the animate and inanimate worlds appears deeper than ever.

PHYSICISTS SAY THE multiverse saddles us with a paradox. Multiverse cosmology builds on cosmic inflation, the idea that the universe underwent a short burst of rapid expansion in its earliest stages. Inflationary theory has had a wealth of observational support for some time but has the inconvenient tendency to generate not one but a great many universes. And because it doesn't say which one we should be in—it lacks this *information*—the theory loses much of its ability to predict what we should see. This is a paradox. On the one hand, our best theory of the early universe suggests we live in a multiverse. At the same time, the multiverse destroys much of the predictive power of this theory.

As a matter of fact, this wasn't the first time Stephen was confronted with a mystifying paradox. Back in 1977 he put his finger on a similar conundrum having to do with the fate of black holes. Einstein's general relativity theory predicts that nearly all information about anything falling into a black hole remains forever hidden inside. But Stephen discovered that quantum theory adds a paradoxical twist to this story. He found that quantum processes near the surface of a black hole cause the hole to radiate a slight but steady stream of particles, including particles of light. This radiation—now known as Hawking radiation—is too faint to be detected physically, but even its mere existence is inherently problematic.[19] The reason is that if black holes radiate energy, then they must shrink and eventually disappear. What happens to the huge amount of information hidden inside when a black hole radiates its last ounce of mass? Stephen's calculations indicated that this information would be lost forever. Black holes, he argued, are the ultimate trash cans. However, this scenario contradicts a basic principle of quantum theory that dictates that physical processes can transform and scramble information but never irreversibly obliterate information. Once again we arrive at a paradox: Quantum processes cause black holes to radiate and lose information, yet quantum theory says this is impossible.

The paradoxes to do with the life cycle of black holes and with our place in the multiverse became two of the most vexing and hotly debated physics puzzles of the last decades. They are concerned with the nature and fate of information in physics and thus strike at the heart of the question as to what physical theories are ultimately about. Both paradoxes emerge in the context of so-called semiclassical gravity, a theoretical description of gravity pioneered by Stephen and his Cambridge gang in the mid-1970s, based on an amalgam of classical and quantum thinking. The paradoxes come about when one applies such semiclassical thinking either over exceedingly long timescales (in the case of black holes) or out to exceedingly large distances (in the case of the multiverse). Together they embody the profound difficulties that arise when we try to get the two pillars of twentieth-century physics, relativity and quantum theory, to work in harmony. In this role they have served as mind-bending thought experiments, with which theorists have extrapolated their semiclassical thinking about gravity to the extreme to see where and how exactly it would break down.

Thought experiments were always a favorite of Stephen's. Having renounced philosophy, Stephen loved to experiment with some of the deep philosophical questions—whether time had a beginning, whether causality was fundamental, and, most ambitious of all, how we as "observers" fit into the cosmic scheme. And he did so by framing these questions as clever experiments in theoretical physics. Three of Stephen's landmark discoveries all resulted from ingenious, carefully configured thought experiments. The first of these was his series of big bang singularity theorems in classical gravity; second, his 1974 discovery in semiclassical gravity that black holes radiate; and third, his no-boundary proposal, also in semiclassical gravity, for the origin of the universe.

Now, while one could argue that the black hole paradox is of academic interest only—the fine details of Hawking radiation are unlikely ever to be measurable—the multiverse paradox bears directly on our cosmological observations. At the heart of the paradox lies the fraught relation in modern cosmology between the living world and *observership,* and the physical universe. The multiverse paradox became a beacon in Hawking's quest to reenvision this relationship by developing a fully quantum perspective on the cosmos. His final theory of the universe, thoroughly quantum, redraws the basic foundations of cosmology and is Hawking's fourth great contribution to physics. The grand thought experiment that lies behind the theory had in some sense been five centuries in the making. Carrying it out would be our voyage.

FIGURE 7(B). *Stephen (left) and the author (right) in the year 2001, shortly after embarking on their journey, in the Brussels bar À La Mort Subite.*

CHAPTER 2

DAY WITHOUT YESTERDAY

L'espace-temps nous apparaît semblable à une coupe conique. On progresse vers le futur en suivant les génératrices du cône vers le bord extérieur du verre. On fait le tour de l'espace en parcourant un cercle normalement aux génératrices. Lorsqu'on remonte par la pensée le cours du temps, on s'approche du fond de la coupe, on s'approche de cette instant unique, qui n'avait pas d'hier parce qu'hier, il n'y avait pas d'espace.

We can compare spacetime to an open, conic cup. We move forward in time by following the cone upward to the top. We move through space by going around in circles. If we imagine going back in time, we reach the bottom of the cup. This is the first instant, the now which has no yesterday, because, yesterday, there was no space.

—Georges Lemaître, *L'hypothèse de l'atome primitif*

IN AN INTERVIEW[1] BROADCAST ON THE BELGIAN RADIO NETWORK IN April 1957 to commemorate the second anniversary of Albert Einstein's passing, Georges Lemaître recalled Einstein's reaction when he first told him of his discovery that the universe expands. This was in October 1927 in Brussels, where many of the world's most eminent physicists had gathered for the Fifth Solvay Council on Physics. The thirty-three-year-old priest-astronomer wasn't among the participants of the council but approached Einstein in the margins of the meeting. However, when he propounded that his general theory of relativity predicted that space expands and that we should therefore see galaxies moving away from us, Einstein balked. "After a few favorable technical remarks, he [Einstein] concluded by saying that from a physical viewpoint this seemed to him 'abominable,'" Lemaître recalled in the interview.

Undeterred, Lemaître took his own findings seriously and thought that the expansion meant that the universe must have had a beginning in what he called a *primeval atom*, a tiny speck of staggering density whose gradual disintegration would have created matter, space, and time.

Why did Einstein vehemently object to the idea of the universe having a beginning? Because he felt this would destroy the very foundations of physics. He thought Lemaître's primeval atom, or any other kind of big bang origin, would be an entry point for God to interfere in the workings of nature. On long walks they took in the early 1930s, Einstein pressed Lemaître to find a way to avoid a beginning, because "this reminds me too much of the Christian dogma of creation." He felt that if cosmological theory gave the universe a birth certificate, it would forever have to remain silent on who or what had issued it, removing all hope of understanding the universe at its most fundamental level on the basis of science.

In vain the Belgian abbé attempted to assuage Einstein, arguing that "the hypothesis of the primeval atom is the antithesis of the supernatural creation of the world."[2] As a matter of fact, Lemaître saw the origin of the universe as a wonderful opportunity to expand the reach of the natural sciences.

Einstein versus Lemaître on the ultimate cause of the universe's expansion went to the heart of the mystery of its apparent design. Their debate was in many ways a precursor of Linde versus Hawking seventy years later. What did Lemaître have in mind when he envisaged the big bang as the "antithesis of supernatural creation"? To understand that, we need to take a closer look at the ideas of both of these scientists.

THE THEORETICAL FOUNDATIONS of modern cosmology rest on Einstein's theory of relativity. This takes us back to the turn of the twentieth century, to a time when physicists had Newton's laws of gravity and motion and James Maxwell's theory of electricity, magnetism, and light, which, together with the theory of heat, underpinned the industrial revolution. The worldview emerging from these nineteenth-century physical theories was in line with our intuitive picture of reality, involving particles and fields, propagating through a fixed space and guided by a universal clock—a cosmic Big Ben one might say. It isn't surprising, therefore, that physicists thought they were converging on a definitive description of nature and that physics would soon be complete.

In the year 1900, however, the Irish-Scottish physicist William Thomson, one of the giants of classical nineteenth-century physics better known as Lord Kelvin, noted "two dark clouds on the horizon."[3] One of the clouds Kelvin identified had to do with the motion of light through the ether and the other with the amount of radiation emitted by hot objects. Still, most physicists felt that these were mere details to be wrapped up and that the edifice of physical theory was solid and sound.

Within a decade, however, this edifice had crumbled. The resolution of Kelvin's "details" unleashed two full-scale revolutions, of relativity and quantum mechanics. What's more, each of these revolutions took physics in a radically different direction, casting a new cloud that still hovers over the frontier of physics today: the problem of how the macro and the micro worlds fit together.

What exactly was it about light that rocked the foundations of nineteenth-century physics? Speed. Careful experiments showed that light always moves at 186,282 miles per second, regardless of the observer's motion relative to a light ray's source. Obviously, that fact didn't match up with everyday experience: If you're traveling on a moving train and you measure its speed, you will clearly obtain a different value (zero) compared to when you measure its speed while standing outside. It also ran against ingrained nineteenth-century thinking. Light waves were thought to be carried by the ether, a mysterious space-filling medium. If that were the case, however, observers moving at different velocities relative to the ether should see light waves pass by at different speeds. But experiments showed otherwise, and that was reason enough for Albert Einstein, working as a clerk in a Swiss patent office, to cast doubt on the ether's existence.

Einstein understood that if light is always observed to travel at the same speed, then observers moving relative to each other must have different notions of distance and time. After all, speed is a measure of distance traveled divided by the duration of the journey. According to Einstein, instead of a cosmic Big Ben we all carry our own clock, and though all our clocks are equally accurate, when we move relative to one another, they will tick at slightly different rates, measuring different amounts of time between the same two events. The same is true for distance; one observer's yardstick can differ slightly from another's. There just aren't universally valid measures for duration and distance. This was the crux of Einstein's 1905 theory of special relativity. The term "relativity" here refers precisely to this revolutionary idea that notions of space, time, and simultaneity aren't objective facts but always tied to the view of a given observer.

One may wonder where the differences in distance measured by one observer relative to another go. Are they simply gone? Not quite. They are transformed into an amount of time. You see, motion through space gets mixed with motion through time in Einstein's relativistic universe. When I look at my sister's parked sports car I find that all its motion is through time. But if she speeds away, a tiny bit of its motion through time is channeled into motion through space. My sister's clock will run a tad more slowly than mine. And while this doesn't quite make her the "young lady from Bright," it does lead to a slight temporal mismatch upon her return. The maximum speed is reached when motion through time is fully diverted into motion through space. That is the speed of light—a cosmic speed limit. Loosely speaking, moving at the speed of light through space leaves nothing for traveling through time. If a particle of light had a wristwatch, it wouldn't tick at all.

With these insights, Einstein's theory broke away from the deeply entrenched Newtonian way of looking at the world, in which space was a fixed stage wherein all events played out and time was a straightforward arrow progressing steadily and universally from the infinite past to the infinite future. In Newton's thinking, nothing could ever affect the unbendable nature of space and the linear flow of time. Also, time and space were not interconnected. According to Newton, time had always existed and always would, independent of any space that might or might not exist.

Einstein's special relativity theory challenged all this by forging an intimate relation between space and time. In 1908, the German mathematician Hermann Minkowski, who had been one of Einstein's teachers at the Zurich Polytechnic, completed Einstein's reconceptualization of space and time and famously declared that "henceforth space as such and time as such shall recede to the shadows and only a kind of union of the two shall retain significance."[4] Minkowski melded the three dimensions of

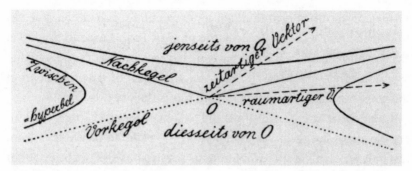

FIGURE 8. *Hermann Minkowski's first diagram unifying space and time into space-time, from his 1908 book,* Raum und Zeit (Space and Time). *Time and one dimension of space are indicated by dashed arrows or "vectors." One arrow points in the time direction ("zeitartiger vector") and one in the space direction ("raumartiger vector"). An observer is located at the point O. The region of spacetime in her future ("jenseits von O") is bounded by the "Nachkegel" and her past ("diesseits von O") is bounded by the "Vorkegel," which are the observer's future and past light cones, respectively.*

space and the one time direction into a single four-dimensional entity: *spacetime.*

In order to picture this four-dimensional union we usually suppress one or two of the three space dimensions and draw the remaining one(s) against the dimension of time in a spacetime diagram. Figure 8 shows Minkowski's very first drawing of spacetime, in which he retained only one dimension of space, running horizontally, and time going vertically. The structure reveals how special relativity redefines our relationship with the universe. If we sit at the point marked O, for observer, then signals traveling at the speed of light that reach us from opposite directions in the past, and signals radiated from O into the future, trace out two lines that cross each other at O and divide spacetime into four distinct parts. The observer's past is the triangular region of spacetime bounded by the trajectories of light rays coming in at O. This region contains all events that have occurred and can affect what the observer sees. The observer's future is the triangular region bounded by light rays leaving O, which contains everything that the observer can influence. Later we will encounter spacetime diagrams that include a second space dimension in the horizontal plane. In such diagrams the paths of past and future light rays at every point trace out two cones touching at their tip at that point and opening up in opposite directions. This light cone structure at every point of space-time is the very essence of relativistic physics. People used to think that the past and the future were simply glued to each other at the present. But special relativity teaches us that for you, the observer, they touch only at the point marking your particular location in the universe.

In Newton's world of distinct and absolute time and space—and no cosmic speed limit—it was thought that we could, in principle at least, instantly access all of space. In Einstein's relativistic world we begin to appreciate how little of it is accessible. The observable universe is limited, both in space and in time, to the region within our past light cone. And given that only 13.8 billion years have elapsed since the big bang, this means there is a *cosmological horizon,* a limiting distance beyond which all happenings in the universe—or multiverse—lie truly out of reach, no matter how much telescope technology ever advances.

Even within our cosmological horizon we can gather information about limited patches of spacetime only. Figure 9 shows the regions within the past light cone of an earthbound observer that are directly accessible. First, astronomical observations of light provide us with information about the near-surface region of the light cone taking us back more than 13 billion years into the past. Second, observations of terrestrial fossils, cosmic particles, and other space debris allow us to look back about 4.6 billion years into the local interior of our past light cone. But there are vast regions in between (shaded lightly in figure 9) to which we have no direct access.

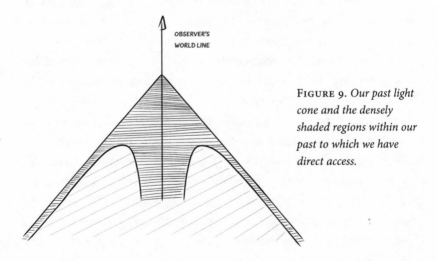

OBSERVER'S
WORLD LINE

FIGURE 9. *Our past light cone and the densely shaded regions within our past to which we have direct access.*

. . .

In 1907, Einstein set out to rethink Newton's law of universal gravitation in order to bring our description of gravity in line with his new relativistic vision of spacetime. This would prove quite a challenge, a mathematical expedition he later described as "a long and lonely journey through the desert, searching in the dark for a truth that one feels but cannot express."[5] But it paid off. In November 1915, in the dark days of World War I, Einstein could finally put forward his theory of general relativity, a new theory of gravity, consistent with his special relativity theory of spacetime, that would become his most sweeping scientific accomplishment.

General relativity describes gravity in geometric terms—the geometry, in fact, of spacetime itself.[6] The theory conceives of gravity as a manifestation of the curving and bending of the fabric of spacetime by mass and energy. For example, the theory holds that Earth moves around the sun not because there is a mysterious force acting over that vast distance, somehow pulling upon Earth, but because the mass of the sun slightly warps the shape of space in its vicinity. This warping creates a sort of valley that guides Earth (and the other planets) into nearly elliptical orbits around the sun. We can't see this valley but we feel it—it's gravity! Likewise, according to Einstein, you are kept with your feet on the ground because the mass of Earth creates a slight dent in the shape of space in which your body tries to slide down, as it were, leading you to feel an upward pressure on your feet. It is that same dent that keeps satellites like the International Space Station and the moon nicely in orbit around our planet.

And it's not only space that curves but also time, a phenomenon exploited—and greatly exaggerated—by the directors of such movies as *Interstellar*. When Joseph Cooper and his crew returned to their spaceship after their brief stay on Miller's planet, they found that Romilly, the crew member who had stayed behind, had aged more than twenty-three years. Apparently the huge mass of the black hole near Miller's planet had caused time to elapse more slowly for the visiting crew.

The sheer power of Einstein's general theory of relativity stems from the fact that it encapsulates this wonderful dialogue between matter and energy and the shape of spacetime in a mathematical equation:

$$G_{\mu\nu} = \frac{8\pi G}{c^4} T_{\mu\nu}$$

This equation isn't difficult to read. On the right-hand side we have all the matter and energy in a region of spacetime, denoted by $T_{\mu\nu}$. The left side describes the geometry, $G_{\mu\nu}$, of that region. The equal sign in the middle is where the magic happens: It tells, with mathematical precision, how the geometry of spacetime on the left ($G_{\mu\nu}$) is tied to a given configuration of matter and energy ($T_{\mu\nu}$) on the right, and this relationship, Einstein's theory says, is what we experience as gravity. Hence gravity doesn't enter Einstein's theory as an independent force. It rather *emerges* from the interplay between matter and the shape of spacetime. As the American physicist John Archibald Wheeler put it, "Matter tells spacetime how to curve. Spacetime tells matter how to move."[7]

In short, general relativity breathes life into spacetime. The theory transforms spacetime from Newton's immutable stage beyond our understanding into a flexible, physical field. Incidentally, the concept of fields in physics, invisible space-filling substances, goes back to the brilliant nineteenth-century Scottish experimenter Michael Faraday. Shortly after, Maxwell made use of fields to formulate his theory of electromagnetism. The magnetic field through which a magnet exerts its influence is probably the best-known example of a physical field indeed. Today, physicists use fields not only to describe forces but also species of particles. Roughly, we think of particles as dense nuggets of their underlying space-filling field. Einstein's genius was to identify spacetime itself as the physical field responsible for gravity.

It wasn't long before support for general relativity began rolling in. The first evidence came from within the solar system and had to do with the path of the planet Mercury. When in the mid-nineteenth century Le Verrier had pointed astronomers to the planet Neptune, he had also noticed that Mercury's orbit slightly deviated from what Newton's law of gravity predicted. Not surprisingly, Le Verrier suggested that Mercury's trajectory might be influenced by another planet, closer still to the sun, and he even had a name for it—Vulcan. But Vulcan was never found. So in 1915 Einstein set out to recalculate Mercury's orbit on the basis of his new theory of gravity and found that it perfectly accounted for the Mercury anomaly, a discovery he called the strongest emotional experience in his life—"as if Nature had spoken."[8]

But the real breakthrough of general relativity came in 1919 when the British astronomer Sir Arthur Eddington sailed out to the Portuguese island of Principe, off the coast of West Africa, to measure the positions of stars during a total solar eclipse. If Einstein were right and mass curves spacetime, then starlight passing near a massive object like the sun shouldn't travel in a straight line but be deflected, causing a slight shift of the star's position in the sky. Strikingly, this was precisely what Eddington and his team found: The stars had moved. *The New York Times* reported on Eddington's observations with the sensational headline LIGHTS ALL ASKEW IN THE HEAVENS, MEN OF SCIENCE MORE OR LESS AGOG, propelling Einstein to international fame as the genius who had dethroned Newton.[9] Newton's laws, once regarded as definitive truths, were shown to be provisional and approximate. That a British astronomer had tested a German physicist's theory was even heralded as an act of reconciliation between both countries so recently embattled in World War I.

The bending of light around the sun is tiny—a few arc seconds—because the sun's gravitational field is weak by astronomical standards. But almost exactly a hundred years later, in the spring of 2019, the world's front pages carried a spectacular smiley image displaying the deflection of light in its most extreme form. In a modern-day version of Eddington's expedition, an international team of astronomers had created a virtual Earth-sized telescope, the Event Horizon Telescope, comprising eight radio dishes across the globe, from Greenland to Antarctica, operating meticulously together to achieve a spatial resolution capable of spotting a tennis ball on the moon. When the astronomers zoomed in with their Event Horizon Telescope and all its resolving power on the very center of Messier 87—a large galaxy in the Virgo cluster, about fifty-five million light-years away—and then digitally stitched together the pixels, a dark disk emerged, surrounded by a halo of light—the hallmark of the shadow of a giant black hole absorbing matter.

The dark disk in figure 10 indicates that there is a central region where the spacetime warping is so enormously strong that light rays straying there aren't merely deflected but remain trapped inside. The ring of light surrounding it arises from matter and gas heating up while disappearing

FIGURE 10. *This first image of a black hole enthralled the world when the Event Horizon Telescope produced it in 2019. The central "shadow" is no larger than our solar system but contains a mass of roughly 6.5 billion suns. It is located in the central nugget of the galaxy Messier 87, about fifty-five million light-years away. The halo of light comes from matter falling into the black hole, while the shadow is where the space warp is so powerful that all light gets sucked in.*

into the black hole. This particular hole is spinning in such a way that the light reaching us from underneath the black disk gets a boost in energy, making the bottom part shine brighter. With a mass of 6.5 billion suns, compressed in a region of roughly the size of the solar system, this is one of the heavier black holes in our cosmic neighborhood.

GENERAL RELATIVITY THEORY had actually predicted that black holes should exist. A mere few months after Einstein's landmark publication, the German astronomer Karl Schwarzschild found the first solution to the theory's defining equation (p. 42) that described the strongly curved geometry outside an extraordinarily dense, perfectly spherical mass M. Since Schwarzschild was serving on the Russian front during World War I at the time, he wrote his solution on a postcard he sent to Einstein in Berlin. Einstein was delighted, of course, and enthusiastically presented the solution before the Prussian Academy.

Schwarzschild's geometry contained a most peculiar surface, located a short distance $2GM/c^2$ from the center of the mass.[10] At this surface space and time appeared to swap roles. For years there was a great deal of confusion about this. Einstein thought this was a mathematical oddity of the solution without physical significance. Schwarzschild himself thought space and time somehow ended at this surface.

But the fog surrounding Schwarzschild's geometry began to clear in the 1930s,[11] when it became apparent that the solution describes the final shape of spacetime after the complete gravitational collapse of a large, perfectly spherical star when it runs out of fuel and dies.[12] Of course real stars aren't perfectly spherical, so most physicists remained skeptical whether such "gravitationally collapsed stars," or black holes, really existed. It would take until the renaissance of general relativity in the 1960s, spurred on by the work of Roger Penrose, before the physical reality of gravitationally collapsed stars finally began to sink in, and Wheeler coined the term *black hole* to describe them. Penrose, a pure mathematician working at Birkbeck College in London, introduced a whole new set of clever tools to handle the complicated geometries of general relativity and proved that all sufficiently massive stars, regardless of their initial shape or composition, collapse to a black hole at the end of their life span. This meant that, far from a mathematical eccentricity, black holes should be an integral part of the cosmic ecosystem. In a 1969 paper, Penrose wrote: "I only wish to make a plea for black holes to be taken seriously and their consequences to be explored in full detail. For who is to say that they cannot play some important part in the shaping of observed phenomena?"[13] These proved prescient remarks. Astronomical observations over the next few decades would gradually strengthen the case for black holes, culminating in the first shadowy images of these enigmatic objects in 2019. Fifty-five years after his prediction that black holes should be ubiquitous in the universe, Penrose shared the 2020 Nobel Prize in Physics for what was initially a purely theoretical discovery.

Penrose's Nobel Prize–winning 1965 paper[14] is just three pages long and has few equations, but it contains a fascinating, Leonardo-like drawing of the formation of a black hole out of a collapsing star (see figure 11). Penrose's spacetime diagram shows two spatial dimensions and how they intermingle with the dimension of time. We can see that far away from the object, the future light cones open up toward both sides, meaning that beams of light can be directed either toward or away from the star—as one expects. Near the collapsing star, the mass of the star curves space, caus-

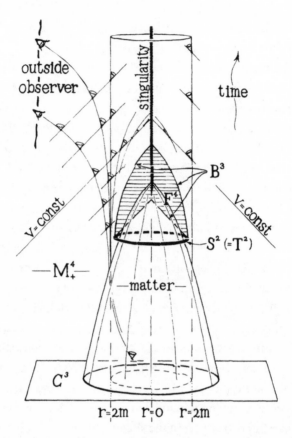

FIGURE 11. *Roger Penrose's 1965 illustration of a star collapsing to form a black hole. As the star shrinks, a curious surface appears in the empty space around it, shown as the black ring in the center of the figure. On that surface, even light can't move away from the star. Penrose demonstrated on purely mathematical grounds that, regardless of its shape, the emergence of such a light-trapping surface signals the inevitable forma- tion of a black hole, with a singularity at the center surrounded by a cylindrical event horizon. Inside the black hole, the extreme tilting of the future light cones means one must keep moving toward the singularity. However, that tilting also means that an ob- server outside never quite sees the final stages of the collapse, let alone the singularity inside the black hole.*

ing the light cones to bend inward. As the collapse proceeds, a special surface appears at which the cones get bent so much that even outward-directed light rays, moving out at the speed of light, remain hovering at a constant distance from the center of the star. And since nothing can travel faster than light, neither can anything else escape its gravitational pull.

The collapsing star has created a region of spacetime that is completely disconnected from the rest of the universe—a black hole.

The surface that separates the no-escape zone inside from the rest of the universe is the peculiar surface in Schwarzschild's geometry that caused so much confusion in the early days of general relativity. Today it is known as the *event horizon* of a black hole. It corresponds roughly to the edge of the dark disk in figure 10. The event horizon acts like a one-way membrane through which matter, light, and information can enter, but nothing can leave. Black holes are the ultimate escape rooms indeed.

Few physicists believe there is much to see or feel at the event horizon of a large black hole, but the horizon is of huge significance to the causal structure of black holes. Inside the horizon surface, space and time in some sense switch identities. If an intrepid astronaut were to venture inside the horizon of a black hole, the ever-increasing tilting of the light cones means he would necessarily have to keep moving toward the center. That is, the radial dimension of space inside acquires the properties of a time dimension, a direction in which one can't stop or reverse but must move forward. The spacetime singularity of infinite curvature that awaits in the center isn't really a place in space but more like a moment in time—the last one.

This singularity with its infinite warping is where (when) the Einstein equation loses its predictive power. The theory of general relativity breaks down at spacetime singularities. This is puzzling. How was Penrose able to prove that the gravitational collapse of a large star produces a singularity when the theoretical framework on which he relied doesn't hold at singularities? The ingenuity of Penrose's strategy was to identify a point of no return in gravitational collapse, the formation of what he called a *trapped surface* from where even light can't move away from the star. Penrose showed that once a trapped surface forms, further collapse to a singularity becomes unavoidable. His mathematical tricks were so powerful that they allowed him to predict the outcome, despite not being able to follow the collapse of a real star through to its completion.

NOW, WHAT HAPPENS when two black holes enter each other's sphere of influence and begin to circle each other? General relativity predicts that this interaction generates gravitational waves, oscillating disturbances of spacetime that propagate across the universe at the speed of light. This is just the Einstein equation at work: Two black holes orbiting each other form a periodically changing configuration of masses to which, the Einstein equation says, spacetime responds with its own periodic disturbances. These ripples are gravitational waves.

As ripples of geometry, gravitational waves carry massive amounts of energy. This drains the energy from the system of orbiting black holes, causing them to spiral inward and eventually merge to form a single, bigger black hole. Such mergers are by far the most violent events in the universe. A single collision of two black holes can generate a burst of gravitational waves more powerful than the combined power of all light radiated by all the stars in the observable universe. Nonetheless, the waves of geometry generated when black holes collide have an extremely small size, because the fabric of spacetime is extraordinary stiff.[15] This is why bursts of gravitational waves, despite their enormous power, are very difficult to detect.

Furthermore, since there are no particles involved, a burst of gravitational waves passing through our planet is a supremely elusive event. Except for momentarily causing yardsticks to stretch and shrink a tiny bit, and clocks slightly to speed up and slow down, gravitational waves zip through planets as if they are carrying an invisibility cloak. In order to detect the fleeting variations caused by gravitational ripples, you need a yardstick several miles long, capable of measuring that distance to a precision considerably better than the width of a single proton. This sounds impossible. In a stunning feat of engineering, however, two groups of scientists, the LIGO team in the United States and the Virgo team in Europe, have done exactly that. By employing lasers and a great deal of sophisticated engineering to monitor the length of three pairs of several-miles-long vacuum tubes placed in an L-shaped configuration, at three widely separated locations on the surface of Earth, both teams have set an ingenious trap for gravitational waves passing through our planet. And on September 14, 2015, after years of waiting and listening, the legs of the two LIGO L's suddenly began vibrating, ever so slightly at first, gradually faster and more strongly, until, after a fraction of a second, the vibrations ebbed away again. Using Einstein's theory, physicists were able to trace this brief pattern of vibrations to a gravitational wave burst generated in the inward spiral and merger, more than a billion years ago, of a pair of black holes of around thirty solar masses each. Five years on, nearly a hundred such gravity wave bursts had been detected, revealing that black holes are an integral part of the cosmic ecosystem indeed, just as Penrose predicted.

The observational discovery of gravitational waves confirms the last of the great predictions enshrined in Einstein's general relativity. In many ways, it marks a coming of age of the theory, for it signals as much a new beginning as the closure of an era. What started out as an abstract mathematical equation describing space, time, and gravity has developed, with the observation of gravitational waves, into an entirely new way of looking at the universe. More than four centuries after Galileo first pointed a telescope to the stars, it is as if astronomers have grown a new sense that they can use to unlock the universe's dark side, dominated by black holes, dark matter, and dark energy. The gravitational-wave observatories that are currently operational across the globe explore the universe by sensing the geometry of spacetime itself, capturing minute vibrations of the field that Einstein birthed more than a century ago.

BACK IN THE pioneering days of general relativity, Einstein was quick to realize that his theory might embody a radical new vision of the cosmos as a whole. In 1917 he wrote to the eminent Dutch astronomer Willem de Sitter in Leiden: "I want to settle the question of whether the basic idea of relativity can be followed through to its conclusion, and determine the shape of the universe as a whole."[16]

Einstein proposed that the global shape of space is like a three-dimensional version of the surface of a sphere—a so-called *hypersphere*. Hyperspheres are difficult to imagine because we tend to think of curved spaces as two-dimensional surfaces embedded in normal three-dimensional Euclidean space. But this embedding of surfaces in a bigger space is only there to provide some comfort for the eye. Nineteenth-century mathematicians had already shown that all geometrical properties of curved surfaces—things like straight lines and angles and so on—can be defined intrinsically, without referring to anything above or below the surface.[17] Likewise, the curved shape of a three-dimensional hypersphere needs no external reference point. It is just what it is: a hypersphere.

Like the surface of a sphere, a three-dimensional hypersphere has no center or boundary. Space looks the same wherever you are in a hypersphere. Yet the total volume of space is finite in Einstein's universe. Hence, much as Earth has a finite surface, the number of distinct places in a hyperspherical universe is limited. In effect, if you kept going in a straight line in Einstein's universe, you'd eventually return home from the opposite direction from which you had left, just as you can fly around the globe by always aiming straight ahead. What's more, nothing would have changed, for Einstein engineered his universe to be unchanging in time. To achieve this outcome, he even added an extra term to his equation, which he called the cosmological term and denoted with the Greek letter λ, known today as the cosmological constant. Einstein's λ term describes a dark energy of space that manifests itself on the very largest scales of the universe where it gives rise to some sort of antigravity, or cosmic repulsion. Einstein saw that for a hypersphere of a special size, the attractive gravitational pull of all matter and the repulsion induced by the λ term can be perfectly balanced, yielding a universe that neither expands nor contracts and that exists in all past and future eternity. This was the cosmos he was after and the only one, he thought, in line with the deeper physical meaning of his theory.

Einstein's feat to capture the entire cosmos by means of a single equation demonstrated vividly that general relativity could take us where Newton's laws could not. His static, hyperspherical spacetime relates the universe's overall shape and size to the amount of matter and dark energy it contains, showing that his theory really had the potential to provide fantastic answers to age-old questions. With his treatment of the universe as a whole, Einstein in some sense folded the outer sphere of the ancient world models into the realm of modern science. Even though his model universe turned out to be widely off the mark, his pioneering explorations marked the birth of modern, relativistic cosmology.

IT WOULD BE ten more years, however, before Lemaître began to discern that the true cosmological implications of relativity went far beyond what Einstein and everyone else had imagined.

Lemaître was an interesting, amiable figure.[18] Born in 1894 in the city of Charleroi in the south of Belgium, he had to give up his university engineering course for military service in World War I. When Germany invaded Belgium in August 1914, young Georges volunteered to serve in the infantry of the Belgian army and took part in the Battle of the Yser, near the border with France. This conflict dragged on for two months, until the Belgians flooded the area to halt the German advance. In quieter periods, it is said that Lemaître sought to relax in the trenches by reading the physics classics, including Poincaré's *Leçons sur les Hypothèses Cosmogoniques*. Family legend has it that he incurred the wrath of one army instructor when he pointed out a mathematical error in the military ballistics manual.

Responding to a double vocation, after the war Lemaître enrolled at the Catholic University of Louvain to study physics and at the seminary in Malines, where he was granted special permission by Cardinal Mercier to study Einstein's new theory of relativity. In 1923, clerical collar in place, he crossed the English Channel to work with Eddington at the Cambridge Observatory.

An avid reader of philosophy as well as physics, Lemaître may well have been inspired by the vision of the eighteenth-century Scottish philosopher David Hume, to carve out a scientific approach at the intersection of mathematical theory and astronomical observations. In his masterpiece, *An Enquiry Concerning Human Understanding*, Hume held that our experiences are the foundation for our knowledge. While recognizing the power of mathematics, Hume cautioned against abstract reasoning detached from the real world: "If we reason a priori, anything may appear able to produce anything. The falling of a pebble may, for aught we know, extinguish the sun; or the wish of a man to control the planets in their orbits." With his emphasis on experience as the basis of all our theories, Hume helped lay the foundations for a practice of science as an inductive process rooted in experiment and in our observations of the universe.

In a similar spirit, Lemaître summarized his position as follows: "Every idea comes from the real world in some way, according to the adage 'Nihil est in intellectu nisi prius fuerit in sensu.'[19] Certainly, the idea that stems from the fact must go beyond it and follow the natural flow of thought, the fundamental activity of the intellect. Yet this is perhaps one of the most valuable lessons that the strangeness of physics teaches us: This flow must be controlled, it must not lose contact with the facts, it must allow itself to be conditioned by them. Here, as in so many other fields, we must find a happy balance between a dreamy idealism that goes astray and a narrow positivism that remains sterile."[20]

Moving from Cambridge, England, to Cambridge, Massachusetts, to work at the Harvard College Observatory, Lemaître witnessed the closing of "the Great Debate" in Washington in January 1925. The question was whether the spiral nebulae in the sky, which had been known since the Middle Ages, were giant gas clouds within the Milky Way or separate, distant galaxies. Using the world's most powerful telescope at the time, the new Hooker 100-inch telescope on Mount Wilson near Pasadena, the American astronomer Edwin Hubble and his colleagues had resolved portions of two nebulae (Andromeda and Triangulum) into separate stars and then used the characteristic properties of pulsating Cepheid stars in these to estimate their distances.[21] To their amazement, these were approaching one million light-years. Such large distances put them far beyond the edge of our Milky Way, confirming they were separate galaxies indeed. In one stroke, Hubble's observations had made the universe a thousand times larger.

What's more, most nebulae appeared to move away from us. As early as 1913, the gifted astronomer Vesto Slipher, working at the Lowell Observatory[22] near the Grand Canyon, had noticed that the spectra of light from most spiral nebulae were shifted toward light waves of longer wavelengths.[23] Such a shift occurs when one observes light from sources that are receding, a phenomenon known as the Doppler shift. We are familiar with the Doppler shift from sound waves—think about the changing sound of the siren of an ambulance driving by. But the same phenomenon also applies to light waves, where if the light source moves away it gives rise to a reddening of the light's overall color, aptly called a *redshift* in cosmology. By the mid-1920s, Slipher had measured the spectra of no fewer than forty-two spiral nebulae and found that only four were approaching the Milky Way while thirty-eight were moving away, often with enormous velocities, up to 1800 km/sec, much greater than the velocity of any other celestial object known at the time. In hindsight, the nebulae velocities listed in Slipher's tables, such as the one shown in figure 12, were the very earliest indications that the universe expanded.[24]

TABLE I.

RADIAL VELOCITIES OF TWENTY-FIVE SPIRAL NEBULÆ.

Nebula.	Vel.	Nebula.	Vel.
N.G.C. 221	− 300 km.	N.G.C. 4526	+ 580 km.
224	− 300	4565	+1100
598	− 260	4594	+1100
1023	+ 300	4649	+1090
1068	+1100	4736	+ 290
2683	+ 400	4826	+ 150
3031	− 30	5005	+ 900
3115	+ 600	5055	+ 450
3379	+ 780	5194	+ 270
3521	+ 730	5236	+ 500
3623	+ 800	5866	+ 650
3627	+ 650	7331	+ 500
4258	+ 500		

FIGURE 12. *The earliest evidence that the universe expands. Shown are the radial velocities of twenty-five spiral nebulae (galaxies), published by Vesto Slipher in 1917. Negative terms correspond to approaching galaxies, while positive velocities are galaxies that are receding.*

Back in Louvain, in 1925, Lemaître recognized the significance of Slipher's observations. It is said that by then he understood general relativity better than anyone else, Eddington and Einstein included. Lemaître saw that the static universe Einstein had engineered was badly unstable. It looked much like the cosmological equivalent of a needle balanced on its head; give it the slightest nudge, and it starts to move. His stroke of genius then was to abandon the deeply rooted idea of an unchanging universe that is the same for all eternity and to read in general relativity what the theory had been trying to say all along: The universe expands. In tying mass and energy to the shape of spacetime, Einstein's theory inevitably causes space to change over time—not just locally but also *in extenso*, on the scale of the whole cosmos. By designing a static world, Lemaître concluded, Einstein had defied the most dramatic prediction of his own equation in favor of his philosophical prejudices of how the cosmos *should* be. Lemaître's seminal 1927 paper, in which he predicted that space expands, established *the* fundamental link between the theory of general relativity and the behavior of the physical universe as a whole.[25] He himself recalled, with characteristic flippancy, that "I happened to be more mathematician than most astronomers, and more astronomer than most mathematicians."[26]

FIGURE 13. *Georges Lemaître lecturing at the Catholic University of Louvain, Belgium.*

Lemaître understood that an expanding universe is very different from an ordinary explosion. An explosion originates in a particular location. If you consider, say, an exploding star from a distance, space will look very different depending on whether you look toward or away from the star. Not so for an expanding universe. A universe in expansion has no center and no edge, for it is space itself that is stretching. If anything, the expansion is the explosion *of* space. "The nebulae [galaxies] are like microbes on the surface of a balloon," Lemaître elaborated. "When the balloon increases, each microbe realizes that the others withdraw, and it has the impression—but only the impression—of being at the centre." A cartoonesque illustration of Lemaître's metaphor appeared in a Dutch newspaper in 1930 (see insert, plate 2).

While light waves travel from one "microbe" to another, their wavelengths stretch along with the stretching space, steadily reddening the light's color. This makes it seem as if distant galaxies rush away from the Milky Way, even though they don't move. Hence the redshifts of the nebulae spectra aren't quite real Doppler shifts due to the actual motions of galaxies, as Slipher and Hubble thought, but a mere consequence of the swelling of space itself. I have attempted to illustrate this in figure 14. Because I can fit only so many dimensions on a sheet of paper, I have again suppressed two space dimensions, leaving only one, depicted here as a circle. The interior of the circle and the space outside it are not part of this universe, but merely aids to the visualization. So we have a one-dimensional circle that is stretching, the radius of the circle getting bigger in the course of time. We can see that this increases the distances between galaxies.

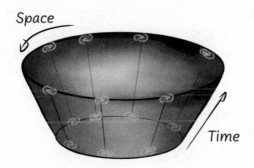

FIGURE 14. *A schematic representation of a one-dimensional, circle-shaped universe that expands in the course of time. The expansion of space makes galaxies rush away from each other, even though they don't actually move. As a consequence of this apparent motion, the light from galaxies we observe appears redshifted.*

The degree of redshift that we observe depends on how long ago—and hence how far away—the light we receive was emitted. Lemaître calculated that if the universe is expanding at a constant rate, then there must be a linear relation between the apparent recession velocity, v, of a galaxy and its distance, r, from us, which he summarized in the infamous equation 23 of his 1927 paper:

$$v = H \, r$$

This relationship says that the apparent velocity v with which galaxies recede should be in proportion to their distance r from us. The proportionality factor H in this relationship is the number that measures the rate of expansion of the universe. Seeking observational corroboration for his prediction, Lemaître looked up Slipher's redshifts and Hubble's (highly uncertain) distance measurements for their sample of forty-two nebulae, and estimated that galaxies are receding about 575 kilometers per second faster for every three million light-years of distance.[27]

THIS DISCOVERY USHERED in the biggest paradigm shift in cosmology since Newton. At the time, however, hardly anyone took notice, and the few remarks that did reach Lemaître weren't encouraging. Lemaître sent a copy of his paper to Eddington, who lost it. Einstein, having fiddled with his theory to get the universe to stand still, refused to reconsider the matter. In effect, he pointed out to Lemaître, during their brief, agitated encounter at the Fifth Solvay Council,[28] that solutions to his equation describing expanding universes had been discovered four years earlier by a young mathematician from St. Petersburg, Alexander Alexandrovich Friedmann,[29] who had since died. To Einstein (and to Friedmann), such solutions were mere mathematical oddities of the relativity theory, without any significance to the real cosmos. A static universe seemed so much more perfect and emotionally pleasing. For all we know, then, with Friedmann dead, Einstein in denial, and Eddington oblivious to Lemaître's discoveries, in the late 1920s only a single person on the planet apprehended what would eventually prove general relativity's most far-reaching prediction.

But Lemaître persevered and set out to study the course of the universe's growth. Working at his home in Louvain—a former brewery—Lemaître traced the evolution of the size of a three-dimensional hypersphere[30] filled with different amounts of matter and dark energy. In the insert, plate 1 shows the range of universes he found, each one expanding and evolving according to the theory of general relativity. These graphs, meticulously calculated on yellow millimeter paper by Lemaître in 1929 or 1930, constitute one of the most remarkable scientific documents of the twentieth century. Epic in the sense of their departure from the prevailing worldview, they literally changed the world.

IN 1929, WITH the world's most powerful telescope on Mount Wilson still at his disposal, Hubble provided strong empirical evidence for the linear distance-velocity relation, to the extent that this relationship—equation

23 in Lemaître's 1927 paper—became known as the *Hubble law*.[31] This was despite the fact that Hubble made no mention of expansion and went to his grave not believing in the relativistic interpretation of his observations.[32] Leaving that aside, the observations were a real tour de force. Hubble was assisted by Milton Humason, a former mule driver and one of the last astronomers to enter the field without a university degree, who went to heroic efforts to capture the faint light of the distant nebulae. It is said that it took Humason three full nights of careful observing to measure the spectrum of a single nebula.

Hubble and Humason's spectacular observations proved the tipping point for relativistic cosmology. In January 1930, Eddington convened a meeting at the Royal Astronomical Society to discuss the matter and—having been reminded of Lemaître's 1927 article—ordered that an English translation be published immediately in the *Monthly Notices*. Confronted with the astronomical evidence, Einstein too conceded the point. In a single stroke he accepted the expansion and discarded the λ term that he had added to his equation to make the universe stay put. He had always had a bad feeling about that term, he said, which he thought was gravely detrimental to the mathematical beauty of his theory. In reference to his newly unencumbered theory, he wrote to the American astronomer Richard Tolman, saying, "This is really incomparably more satisfactory."[33]

Paradoxically, Lemaître held a very different view. He thought Einstein's λ term was a brilliant addition to the theory, not to engineer a static universe, of course (Einstein's motivation), but simply to account for the energy associated with empty space. Eddington agreed with Lemaître on this matter, declaring at one point, "I would rather revert to Newtonian theory than to drop the cosmological constant."[34] Whereas Einstein had added the term on the left side of his equation, reasoning on geometric grounds, Eddington and Lemaître thought of it as part of the energy budget of the universe on the right-hand side. If spacetime is a physical field, they argued, then shouldn't we expect that it comes endowed with its own intrinsic properties? The cosmological constant does exactly that: It fills spacetime with energy and pressure. Just as a bowl of milk contains a certain amount of energy, given by its temperature, the λ term suffuses otherwise empty space with an amount of dark energy and dark pressure determined by the numerical value of the constant λ. "With the λ-term everything happens as though the energy in vacuum would be different from zero," Lemaître wrote.[35]

The antigravity effect of the cosmological constant comes about because the pressure it fills space with is negative. Negative pressure isn't particularly exotic; it's what we often call tension, as in a stretched rubber band. Negative pressure induces "negative gravity," or antigravity, in Einstein's theory, and this is what speeds up the expansion.

Now, when space stretches, its intrinsic properties don't change. You just get more of it. So, unlike the energy of normal matter or radiation, the dark energy of spacetime doesn't dilute as the expansion unfolds and may even become the determining factor in the universe's evolution when space grows large. This is not the case in the hyperspherical universes that correspond to the lower set of curves in Lemaître's iconic graph (see insert, plate 1). In these universes, space has a small dark energy density. As a consequence, gravity is overall attractive and the size of the universe changes much like the trajectory of a baseball in flight: It starts out growing, reaches a maximum before the antigravity of the dark energy builds up and interferes, and then collapses again into a *big crunch*. But if the value of the cosmological constant were larger, it could counter the gravitational pull of matter and dramatically alter the course of cosmological evolution. The path of expansion of universes with enough dark energy transitions from something like the trajectory of a baseball into an accelerating space rocket. This is the kind of behavior Lemaître displays in the upper curves of his diagram.

As a matter of fact, besides thinking about the properties of empty space, Lemaître had a second reason to keep λ in his basket, one that was no less interesting and that I already alluded to in chapter 1. This had everything to do with the habitability of the universe. By carefully adjusting the numerical value of λ, he could obtain a universe with a long era of very slow expansion in which galaxies, stars, and planets could form. This hesitating universe was by far the most biofriendly one that Lemaître came up with: It corresponds to the one curve going nearly horizontally in plate 1. (Even this universe, however, if Lemaître had kept on calculating, would eventually have started accelerating.)

Lemaître and Einstein continued to quarrel about "little lambda" for the rest of their lives. Agreement was never reached. Journalists trailing them on their walks around the Athenaeum at Caltech wrote of the "little lamb" following them everywhere they went. In his later correspondence with Lemaître on the subject, Einstein conceded that if he "could demonstrate that λ were present, this would be very important."[36] This was as far as he ever came to reconsidering the infamous λ term. No fewer than eighty years later, in a truly remarkable development, high-precision astronomical observations of the spectra of exploding stars called supernovae would prove Lemaître right: We do live in a hesitating universe, though its period of hesitation ended a few billion years ago.[37]

BUT PERHAPS THE most mind-boggling "detail" in Lemaître's diagram in plate 1 hides in the bottom left corner, where he wrote "t = 0," the *zero of time*.

You see, Lemaître's original 1927 expanding universe did not have a beginning. Instead, Lemaître had assumed that the universe had evolved slowly and gradually from a near-static state in the infinite past. By 1929, he had realized that this arrangement in the far past was much like Einstein's needle balanced on its head, so he abandoned this scenario in favor of a genuine beginning. Lemaître arrived at the conclusion that the expansion meant that the universe must have had a past that was unimaginably different from the present. "We need a complete revision of our cosmogony," he held, "a fireworks theory of cosmic evolution."[38]

Venturing far beyond where even Einstein's theory could take him, he came to envision the origin of the universe as a super-heavy primeval atom whose spectacular disintegration would yield the vast cosmos we see today. "Standing on a cooled cinder we see the slow fadings of the suns and try to recall the vanished brilliance of the origin of the worlds," he wrote in his monograph *L'hypothèse de l'atome primitif*. In search of fossil remnants of the universe's violent birth, he then took an interest in cosmic rays, which he thought of as hieroglyphs of the ancient fireball. Later in his career, to decipher their trajectories, Lemaître purchased one of the first electronic computing machines, the Burroughs E101, which he had seen at the 1958 World Expo in Brussels, and, with the help of his students, famously carried up to the attic of Louvain's Physics Department, establishing the university's first computing center.[39]

However, whereas the idea of an expanding universe became broadly accepted in the early 1930s, any talk about the universe having a beginning was met with great skepticism. "The notion of a Beginning of the present order of Nature is repugnant to me," Eddington asserted. "As a scientist I simply do not believe that the universe started with a bang. As if something unknown is doing we don't know what."[40]

Einstein too initially dismissed the idea of a beginning. Much like his view of the singularity inside Schwarzschild's spherical black holes, he thought that the zero of time in Lemaître's expanding universes was an oddity of the perfectly symmetric and uniform manner in which they expanded. Since the real universe isn't perfectly uniform, things would miss one another when you ran the expansion backward, he reasoned, replacing the beginning with cycles of contraction and expansion, which he found philosophically much more satisfactory. Lemaître recalled their conversation in 1957: "I met again with Einstein in California, at the Athenaeum in Pasadena. Speaking of his doubts concerning the inevitability, under certain conditions, of the beginning, Einstein proposed a simplified model of a non-spherical universe for which I had no difficulty in calculating the energy tensor, and to show that the loophole of which Einstein had thought [to avoid a beginning] did not work."[41] Lemaître apparently shared Einstein's feelings on the inevitability of a beginning, noting, "From an aesthetic point of view this is unfortunate. A universe that repeatedly expands and contracts has an irresistible poetic charm, reminiscent of the Phoenix of the legends."[42]

However, the universe is what it is. Notwithstanding the philosophical and aesthetic leanings of its pioneers, relativistic cosmology pointed strongly toward a genuine beginning and has done so ever since. That said, the zero of time—Lemaître's day without yesterday—is again a singularity in general relativity, where the spacetime curvature becomes infinite and, as a consequence, the Einstein equation turns silent. So, curiously, the big bang is the cornerstone as well as the Achilles' heel of relativistic cosmology—inevitable yet apparently beyond understanding.

This is a profoundly confusing state of affairs. If time itself began with the big bang, then all questions about what happened before would seem meaningless. Even speculation about what caused the big bang would be out of place, for causes precede effects, which requires some notion of time. This apparent breakdown of basic causality at the origin of time was the core of the matter in the debate that pitted Eddington and Einstein against Lemaître. The former were so reluctant to contemplate a universe beginning because it felt as though a real beginning required a kind of supernatural agency to interfere with the natural course of evolution. This reticence would become all the more poignant when, throughout the century, more and more evidence emerged that the universe originated in a way that is strikingly conducive to the evolution of life. In hindsight, Eddington and Einstein could be forgiven for being suspicious!

EINSTEIN'S AND EDDINGTON'S perspectives on the beginning were steeped in the old determinism going back to Newton, to which Einstein's classical theory of general relativity comports. Within this scheme, any beginning requires initial conditions with the same degree of tuning as the universe that evolves from them. A universe that evolves to become complex late in its evolution requires initial conditions of the same level of complexity early on. A universe that appears specially designed to bring forth life requires initial conditions that encode that same level of bio-friendliness all the way back at the start. This makes it seem as if an "act of God" was involved to set our fine-tuned, biophilic universe in motion.

But Lemaître was a giant step ahead of determinism. He proposed to break the chain of causes and effects by taking a quantum view of the origin and explained his position in "The Beginning of the World from the Point of View of Quantum Theory," published in the journal *Nature* in May 1931.[43] Lemaître's cosmopoetic letter is one of the most audacious scientific texts of the twentieth century. It counts no more than 457 words but can be regarded as the charter of big bang cosmology. In this letter he argues, to my knowledge for the first time, that the relativity and quantum revolutions are profoundly interconnected, that the beginning of the universe should be part of science, governed by physical laws that we can discover, but that these hypothetical laws will involve a mixture of quantum theory with gravity. We must meld together quantum theory and relativity, Lemaître argued, because the latter implies a big bang where the former becomes important. And that unification, he envisioned, will provide such a powerful and deep synthesis that it will integrate the universe's origin within the realm of the natural sciences. These proved prescient thoughts; today physicists are fond of saying that the big bang is the ultimate quantum experiment.

Quantum theory imbues physics with an unavoidable element of indeterminacy and "fuzziness." Lemaître speculated that under the extreme conditions in the earliest stages of the universe, even space and time would become fuzzy and uncertain. "The notions of space and time would altogether fail to have any meaning at the beginning," he wrote in his big bang manifesto. "Instead space and time would only begin to have a sensible meaning when the original 'quantum' had been divided into a sufficient number of quanta," enigmatically adding, "If this suggestion is correct, the beginning of the world happened a little before the beginning of space and time."

But how could quantum indeterminism resolve the causality conundrum that the big bang poses? What Lemaître had in mind was that random quantum jumps could have generated a complex universe from a simple primeval atom. And if the complexity of today's universe were the result of countless frozen accidents in its embryonic evolution rather than the necessary consequence of perfectly orchestrated initial conditions right at the start, wouldn't the whole idea of a beginning be easier to swallow? Contemplating the implications of a quantum origin, Lemaître ended his letter in *Nature* by saying: "Clearly the initial quantum could not conceal in itself the whole course of evolution. The story of the world need not have been written down in the first quantum like the song on a disc of a phonograph. . . . Instead from the same beginning widely different universes could have evolved."

In fact, since the idea of a quantum origin seemed to take the sting out of the origin of time, Lemaître came to see it as a central pillar of his new cosmology, even though he never wrote down a single equation for the primeval atom that could substantiate his vision. The intuitive picture of the beginning that he muses about in his big bang manifesto is one of supreme simplicity. The primeval atom in Lemaître's thinking was like an abstract, undifferentiated, pristine cosmic egg. It makes me think of the Romanian sculptor Constantin Brâncuşi's *The Beginning of the World* (see insert, plate 6).

The British quantum physicist Paul Dirac, an early supporter of the primeval atom hypothesis, went even further and speculated that quantum jumps in the early universe could replace the need for an initial condition altogether. Could it be that causality fades away in a quantum origin, that the mystery of the "first cause" evaporates in a quantum world—our world?

Paul Dirac had arrived in Cambridge as a student in 1923, the same year as Lemaître, and also with the hopes of studying relativity with Eddington. But he was assigned a different track that drew him into the quantum theory of particles, a field of research in which he would acquire a virtually unmatched depth of understanding. Dirac discovered the eponymous equation that unified Einstein's theory of special relativity with quantum theory and predicted the existence of antimatter, earning him the 1933 Nobel Prize. Later he rose to become the fifteenth holder of the Lucasian Chair of Mathematics in Cambridge. Dirac was a curious character, though, notoriously shy and quiet—almost invisible, according to some colleagues. One day in the late 1970s Stephen and his wife, Jane, invited the Diracs over for tea at their house on a Sunday afternoon. Don Page, Stephen's research assistant at the time who lived with them to help out with Stephen's daily care, decided to stick around to listen in on the conversation between these two giants of twentieth-century physics. Apparently neither of them said a word.

The Dirac archives in Tallahassee, Florida, contain a beautiful pencil drawing of Lemaître made by a member of the audience during Lemaître's lecture at the Kapitza Club in Cambridge in 1930. The drawing, shown in figure 15, comes with a caption that says, "But I don't believe in the Finger of God agitating the aether." According to Dirac's recollections, which he wrote down in an accompanying note in 1971, "there was much discussion during Lemaître's lecture about the role of quantum indeterminacy." Both Dirac and Lemaître saw in quantum mechanics a way to disentangle the causal knot created by a deterministic outlook on the beginning, by tracing the roots of much of the universe's complexity today to random quantum jumps in the wake of its birth. These jumps would in a sense turn cosmological evolution into a genuinely creative process.

Taking stock after a whirling decade of discoveries, Dirac returned to Lemaître's primeval atom hypothesis in 1939, in his Scott Prize Lecture at the Royal Society of Edinburgh: "The new [expanding] cosmology will probably turn out to be philosophically even more revolutionary than

FIGURE 15. *This drawing of Georges Lemaître was made by an attendee at a talk he gave in 1930 at Cambridge University. The note underneath makes clear that Lemaître saw no reason for God to interfere with the big bang. He regarded the primeval atom hypothesis as a purely scientific matter, grounded in physical theory and to be verified, ultimately, by astronomical observations. Forty years later, Paul Dirac wrote down the accompanying note on the right.*

relativity or the quantum theory, although at present one can hardly realize its full implications."[44] Seventy years on—and liberated from a few more prejudices—my journey with Stephen would bring to the surface some of these philosophical implications indeed.

. . .

Around 1930 the Abbé Lemaître visited Cambridge and gave a lecture at the Kapitza Club. There was much discussion about the indeterminacy of quantum mechanics. Lemaître emphasized his opinion that he did not believe God influenced directly the course of atomic events.

A member of the audience made this drawing to commemorate this discussion. I do not remember who the artist was. It is quite a good likeness of Lemaître.

P A M Dirac   1st Sept 1971

Back then, observations that could vindicate the primeval atom hypothesis or something like it remained elusive. After its heydays in the early 1930s, cosmology actually became a bit of a scientific backwater, characterized by scant data and grandiose speculation, and cosmologists acquired the dubious reputation of being "often in error but never in doubt."

In effect, in the 1950s, the big bang theory almost slipped from public view. Vocal in his opposition to Lemaître's theory, the British astrophysicist Fred Hoyle had coined the term big bang as a derisory term, during a BBC radio interview in 1949, and painted it as "an irrational process that cannot be described in scientific terms." Hoyle missed no opportunity to portray big bang cosmology as pseudoscience that had been pursued in a concordist endeavor. Echoing Eddington, Hoyle stated, "There can be no

causal explanation, and indeed no explanation of any kind, for the beginning of the universe. The passionate frenzy with which big bang cosmology is clutched to the corporate scientific bosom evidently arises from a
deep-rooted attachment to the first page of Genesis, religious fundamen-
talism at its strongest,"[45] and he recommended, "Whenever the word 'origin' is used, disbelieve everything you are told!"[46]

Working with Hermann Bondi and Thomas Gold, Hoyle proposed a
rival model of the universe, the steady-state theory, which became a serious contender in the 1950s. The steady-state theory held that even though
the universe is always expanding, it maintains a constant average density
because matter is being continuously created to form new galaxies that fill
the spaces opening up as older galaxies move apart. Whereas in big bang
cosmology most matter is created in the primeval heat, in a steady-state
universe the creation of matter is a slow and everlasting process. The universe has no beginning or end in Hoyle's steady state, which is somewhat
like a mini-version of the multiverse, with a constant production of new
galaxies rather than universes.

Meanwhile, however, the tall Russian physicist George Gamow, Gee-
Gee to his friends, had subjected the exotic environment of a hot big bang
to a closer study. Gamow was a colorful character who seems to have had
a knack for encountering people from all walks of life, from Trotsky and
Bukharin to Einstein and Francis Crick—often in memorable circumstances.[47] Gamow grew up in the Ukrainian city of Odessa and studied in
St. Petersburg, where he learned general relativity from Alexander Friedmann. Dismayed by the growing intrusions of the Communist state into
intellectual life, Gamow and his wife attempted to escape from Ukraine by
paddling out from the southern tip of the Crimean Peninsula across the
Black Sea to Turkey. All went well until two days into their sea journey,
when they got caught in a storm that swept them back to Crimea. But the
Gamows didn't give up. In 1933, when Niels Bohr invited Gamow to at-
tend the Seventh Solvay Council in Brussels, they seized the opportunity
to emigrate to the United States.

Gamow was neither a mathematician nor an astronomer but a nuclear
physicist, who imagined the whole universe as a giant nuclear reactor
during the first few minutes of its expansion. Working with Ralph Alpher
and Robert Herman, Gamow envisaged the big bang to be hot, very hot,

and wondered whether the chemical elements out of which we and everything else around us are made were once cooked in this primeval cosmic oven. He reasoned that if the density and the temperature of the primordial universe were so high that even atomic nuclei could not survive, the periodic table would have been, initially, empty except for the very first element, hydrogen, which is just a single-proton particle. The whole universe would have been filled with a superdense hot plasma Gamow called Ylem, after the Greek ύλη for "matter." This plasma would have consisted of freely moving atomic building blocks—electrons, protons, and neutrons—immersed in a heat bath of radiation. But when the universe expanded and cooled, neutrons and protons would have combined to form composite atomic nuclei. First in line for this is deuterium, heavy hydrogen made of one proton and one neutron, which in turn fuses with more protons and neutrons to form helium. Combining the laws of nuclear physics with the expansion of space, Gamow and his team calculated that the window for nuclear fusion in the primordial universe would open up at around one hundred seconds after the big bang and close again a few minutes later, when the expansion would have lowered the temperature to one hundred million degrees, low enough to shut down the cosmic nuclear reactor. They found that this brief window would be enough to convert about one quarter of all protons in the universe into helium nuclei,

FIGURE 16. *George Gamow overwrote the label of this Cointreau bottle with the word YLEM to commemorate his 1948 work with Ralph Alpher on the synthesis of atomic nuclei in the fiery heat of the big bang. The term "Ylem" is a Middle English word that refers to the primordial substance from which all matter was thought to be created.*

together with a few traces of heavier elements like beryllium and lithium. These relative abundances of the light elements predicted by Gamow and his team are in excellent agreement with what astronomers have since measured. Today, this counts as one of the key tests of the hot big bang theory.[48]

But hiding in Gamow's work was an even more momentous prediction— if this were possible. Alpher, Gamow, and Herman realized that the heat liberated in the synthesis of atomic nuclei should still be around today as a sea of leftover radiation filling all of space. After all, where could it go? The universe is all there is. Their calculations showed that billions of years of cosmic expansion would have cooled the heat radiation to a temperature of about 5 Kelvin, or –267 degrees Celsius. Such cold radiation would make the universe shine predominantly in the microwave frequency band of the electromagnetic spectrum. So the universe today—all of space— should be filled with microwaves. This was a monumental discovery: Gamow and his collaborators had identified a fossil relic of the hot big bang era that, furthermore, should be out there for us to see if only we looked into deep space with eyes sensitive to microwaves.

And indeed it is. Hot bodies radiate, and the universe is no exception. The cosmic microwave background radiation, or the CMB for short, was discovered in 1964 in a moment of serendipity by two American physicists, Arno Penzias and Robert Wilson. Unaware of Gamow's work, Penzias and Wilson were calibrating a giant microwave horn antenna at the Bell Telephone Labs in Holmdel, New Jersey, which had initially been constructed to track the Echo balloon satellites, when they discovered a persistent hiss from their antenna that they could not explain. No matter where in the sky they turned their antenna, they encountered the exact same noise, at a wavelength of 7.35 cm, day and night. Talking with local cosmology friends, they soon realized that their antenna was hissing for a good reason: It was picking up the faint relic radiation of the hot big bang—the telegram from the dawn of time Lemaître had first envisioned and Gamow had later identified.

Penzias and Wilson's discovery of the fossil microwave radiation was the shot heard around the world. At last it dawned upon the scientific community that cosmological expansion had real long-term effects, that it meant that the far past was unimaginably different from the present.

This recognition fundamentally transformed the debate about the universe's origin. Almost overnight the ultimate cause of the expansion, that enigma that had pitted Einstein and Lemaître against each other thirty years earlier, moved to center court in theoretical cosmology, and it has remained there ever since.

LEMAÎTRE WAS TOLD about the discovery of the CMB on the seventeenth of June in 1966, a mere three days before his death, in the hospital, where a close friend brought him the news that the fossil relics that proved his theory right had finally been found. *"Je suis content . . . maintenant on a la preuve,"* he reportedly replied.[49]

Now, it may seem odd that the "father of the big bang" was also a Catholic priest. But Lemaître understood how to navigate between Einstein and the pope and took pains to explain why he saw no conflict between the "two paths to truth," science and salvation, that he had decided to follow. In an interview with Duncan Aikman for *The New York Times,* Lemaître paraphrased Galileo on science versus religion,* saying, "Once you realize that the Bible does not purport to be a textbook on science, and once you realize Relativity is irrelevant for salvation, the old conflict between science and religion disappears," adding, "I have too much respect for God to reduce Him to a scientific hypothesis"[50] (see insert, plate 5). It is abundantly clear from his writings that Lemaître didn't experience the slightest conflict between these two spheres. One even discerns in him a certain lightheartedness about it. "It turns out that to search thoroughly for the truth involves a searching of souls as well as of cosmic spectra," he once said.

In the early 1960s, by then Monsignor Lemaître and president of the Pontifical Academy of Sciences, he would strive to advance the academy's goals of fostering excellent science while maintaining a healthy relation with the Church, by respecting meticulously the differences in methodol-

---

* Galileo in 1615 wrote a legendary letter about the relation between science and religion to Christina of Lorraine, grand duchess of Tuscany. In this he quotes a most eminent ecclesiastic, reportedly Cardinal Caesar Baronius, head of the Vatican library: "That the intention of the Holy Ghost is to teach us how one goes to heaven, and not how heaven goes."

ogy and language between science and religion. Far from the concordist interpretations that sought to bring the truths of faith in line with scientific discoveries, Lemaître insisted that science and religion each had its own playing field. Of the primeval atom hypothesis, he said in this regard, "Such a theory remains entirely outside any metaphysical or religious question. It leaves the materialist free to deny any transcendental Being. . . . For the believer, it removes any attempt to [achieve] familiarity with God. It is consonant with the wording of Isaiah speaking of the 'Hidden God,' hidden even in the beginning of creation."[51]

Lemaître's more formal position on these matters was no doubt influenced by his studies at the neo-Thomistic school of philosophy of Cardinal Mercier in Louvain, which embraced modern science but denied it much ontological significance. At Mercier's institute Lemaître learned to differentiate between two levels of existence, between the beginning of the physical world in a temporal sense and metaphysical questions of existence: "We may speak of this event [the disintegration of the primeval atom] as of a beginning. I do not say a creation. Physically everything happens as if it was really a beginning, in the sense that if something has happened before, it has no observable influence on the behavior of our universe. . . . Any pre-existence of our universe has a metaphysical character."[52]

This distinction made it possible—and indeed obvious—for the abbé to regard the study of the physical origin of the universe as an opportunity for the natural sciences, whereas Einstein regarded it as a threat to physical theory. At the core of their scientific debate, therefore, lay different philosophical positions. They seem to have had very different conceptions of what it is, ultimately, that science tries to find out about the world. Lemaître appears to have been supremely cognizant that no matter how abstract, our ability to do science remains rooted in our relationship with the universe. His double vocation inspired him to carefully delineate the boundaries of both the scientific and spiritual spheres. The result was a faith stripped from dogmas and a science rooted in the human condition. At a memorial event in his native village, one of Lemaître's nieces told me that at family gatherings, her cousins liked to challenge Georges and press him on where his primeval atom came from. "Oh, that is God," he would jokingly tell them.

Einstein, by contrast, was an idealist. His discovery of the theory of general relativity constituted an unparalleled tour de force. This achievement strengthened him in his conviction that there is a final theory of immutable mathematical truths out there, lying in wait to be discovered, that dictates how the universe should be. Einstein's fundamentally causal and deterministic attitude on all questions to do with the origin reflects this. However, the stunning prediction of his own relativity theory that the universe originated in a big bang that is also the origin of time severely challenged his position.

In the next chapters I will argue that Lemaître's position would eventually prove a more trustworthy guide to untangle the riddle of design. In effect Einstein versus Lemaître mirrors the distance Hawking would travel seventy years later. The early Hawking adhered to Einstein's position, to the notion that we are discovering objective truths in physics that somehow transcend the physical universe. The story of our journey at a deeper philosophical level is about how and why Stephen broke with the Einsteinian position and came to adopt Lemaître's, and what this entails not only for our conception of the big bang, but also for the future agenda of cosmology.

FIGURE 17. *When we began to work together, Stephen wasn't aware of Lemaître's pioneering work on quantum cosmology. Hence I took him to Lemaître's former office at the Premonstratensian College in Louvain, where I showed him Lemaître's big bang manifesto of 1931.*

CHAPTER 3

COSMOGENESIS

Ich schreite kaum, doch wähn' ich mich schon weit.
Du siehst, mein Sohn, zum Raum wird hier die Zeit.

I hardly move, yet far I seem to have come.
You see, son, here time becomes space.

—RICHARD WAGNER, *Parsifal*

IN HIS MEMOIR, STEPHEN WROTE THAT HE BECAME INTERESTED IN cosmology because he wanted to fathom the depths of understanding. Stephen's insatiable desire to ask ever deeper questions drew him to Cambridge. He arrived in Cambridge in the autumn of 1962 from Oxford, where he had taken the undergraduate physics course. "The prevailing attitude at Oxford at the time was anti-work," he said about this. "To work hard to get a better degree was regarded as the mark of a grey man, the worst epithet in the Oxford vocabulary."[1] When it came to the final examinations at Oxford, Stephen chose to concentrate on problems in theoretical physics because these didn't require much factual knowledge. He attained a borderline first–second grade, and an interview with the examiners was held to judge which one should be awarded. Stephen told them that if he got a first, the top grade, then he would go to Cambridge; otherwise he would stay at Oxford. They gave him a first. In light of Stephen's later achievements, from Oxford's viewpoint, this must have been one of the worst decisions in its eight-hundred-year history.

In Cambridge, Stephen was drawn to Hoyle, the steady-state man, even though his theory had come under serious pressure in the early 1960s.[2] Hoyle was not available, however, and Stephen was assigned to Dennis

Sciama instead. This proved to be a piece of great fortune. Sciama, himself a student of Paul Dirac, was a catalyst, an outstandingly stimulating figure, who had turned Cambridge into the mecca for relativistic cosmology. Plugged in to the major developments in physics worldwide, Sciama made sure his students knew about the latest discoveries. Whenever an interesting paper was published, he would assign one of them to report on it. Whenever an interesting lecture was scheduled in London, he would send them down on the train to hear it. Stephen throve in the interactive, vibrant, and ambitious scientific environment that Sciama created and would later strive to create an equally stimulating environment for his own students.

When Stephen arrived in Trinity Hall, Cambridge, Sciama too clung to the steady-state model of the universe. He set Stephen to work on a variation of it that Hoyle had devised in an attempt to rescue the theory. Soon Stephen found there were infinities in Hoyle's new version that rendered the theory ill defined, and he challenged Hoyle on this point at a meeting of the Royal Society in London in 1964. When Hoyle asked, "How do you know?" Stephen, refusing to be cowed by Britain's foremost astrophysicist, replied, "Because I calculated it!"—an early sign of both his independent spirit and his flair for drama. His analysis of the steady-state theory would become the first chapter of his PhD dissertation.

The final nail in the coffin for steady-state cosmology came a few months later with the discovery of the cosmic microwave background radiation. The existence of this ancient heat showed beyond a doubt that the universe was not in a steady state but had once been fundamentally different—very hot. But did this also mean that it must have had a beginning? Clearly this was now *the* central question for big bang cosmology, and Stephen was ready to dive in.

SCIAMA PUT STEPHEN in touch with Roger Penrose, who had just published his three-page breakthrough paper showing that black holes should be omnipresent in the universe. If the theory of general relativity holds, Penrose had proven, then the gravitational collapse of stars of sufficient mass winds up creating a spacetime singularity that is hidden from the outside world by an event horizon: a black hole.

Stephen soon realized that if he reversed the direction of time in Penrose's mathematical reasoning so that the collapse became an expansion, then he could prove that an expanding universe must have a singularity in the past.[3] Working with Penrose, he went on to derive a series of mathematical theorems that demonstrate that if one traces the history of an expanding universe back in time, to an era well before the birth of the first stars and galaxies and even before the CMB snapshot, one eventually hits a singularity where spacetime bends to a breaking point. Both sides of the Einstein equation become infinite at the initial singularity—where infinite spacetime curvature "equals" infinite matter density—and this means the theory loses all predictive power. It is somewhat like dividing by zero on your calculator; you'd get infinity, and whatever you compute next makes no sense. Singularities are edges to spacetime, really, where the general theory of relativity offers no guidance as to what happens. Indeed the very word "happen" loses meaning at a spacetime singularity.

Penrose had shown that, according to relativity theory, time must end inside black holes. Stephen's time-reversed argument demonstrated that in an expanding universe time must have a beginning. It is not that the big bang singularity sat there, like a cosmic egg, waiting to hatch a universe. The singularity rather signals the birth of time itself. Stephen's theorem demonstrated that the zero of time in the perfectly spherical model universes of Friedmann and Lemaître wasn't an artifact of their simplicity at all but a robust and universal prediction of relativistic cosmology. This was the central result of his 1966 PhD dissertation—and one that later featured in the biopic *The Theory of Everything*. In the abstract of his PhD thesis, Hawking wrote, "Some implications and consequences of the expansion of the universe are examined. . . . Chapter 4 deals with the occurrence of singularities in cosmological models. It is shown that a singularity is inevitable provided that certain very general conditions are satisfied."

This is a striking result. Walking on Earth's surface in places like the Grand Canyon, one can find rocks that are several billion years old. The simplest forms of bacterial life on Earth are about 3.5 billion years old, and our planet itself isn't very much older, approximately 4.6 billion years. The big bang singularity theorem is saying that if we were to go back to a time just three times earlier—13.8 billion years ago—there would be no

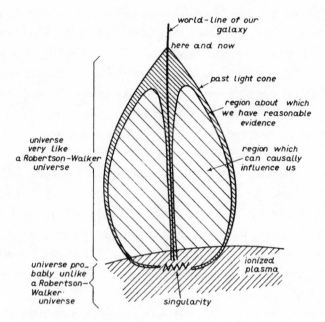

world-line of our galaxy

here and now

past light cone

region about which we have reasonable evidence

universe very like a Robertson-Walker universe

region which can causally influence us

universe pro_ bably unlike a Robertson- Walker universe

ionized plasma

singularity

FIGURE 18. *George Ellis's 1971 drawing of the observable universe and the (finely shaded) parts we can observe in some detail. We are located at the tip, where it says "here and now." Matter causes light rays to converge into the past, bending our past light cone inward and tracing out a pear-shaped region: our past. Since light sets a cosmic speed limit, this region is the only part of the universe that is, in principle, observable to us. According to Stephen Hawking's theorem, the focusing of light into the past means that the past must end in an initial singularity. However, we cannot see directly all the way back to the singularity because photons, particles of light, constantly scatter against everything else in the hot ionized plasma that fills the primeval universe, rendering it opaque.*

time, no space, no anything. Viewed this way, we are rather close to the beginning of everything.

IF STEPHEN HAD still been alive fifty-four years later, he would presumably have shared in the 2020 Nobel Prize in Physics for his singularly important work with Penrose on the beginning and end of time. The picture of our past that emerges from his PhD research is that of a pear-shaped region of spacetime of the kind depicted in figure 18. This wonderful drawing was made by George Ellis,[4] a fellow Sciama student who worked with Stephen on the singularity theorems in the mid-1960s. We are located at the tip of the pear. The surface of the pear is traced out by light rays reaching us from different directions in the sky. The diagram displays the effect of matter on the shape of our past light cone. We can see that the mass of matter causes light rays to deviate from straight lines and to converge as we trace their paths back in time. As a consequence, the straight light cones in figures 8 and 9, which ignore this gravitational focusing effect of matter, are deformed in the real universe and bend inward, creating a pear-shaped surface—our past light cone—that separates the finite patch of spacetime that can influence us, the inside of the pear, from the rest of the universe, which cannot. The crux of Stephen's singularity theorem is that if matter makes past light cones converge in this way, then history *cannot* be extended indefinitely. Instead, one reaches an "edge to moments," a boundary at the bottom of the past where the universe of space and time is no more.

Ellis's drawing is the cosmological analogue of Penrose's iconic illustration of the formation of a black hole shown in figure 11. Comparing the two, we see that an observer's past in cosmology is much like the future inside a massive star—both exist for a finite amount of time only. But there is one crucial difference: While the event horizon of a black hole shields an observer outside the hole from the violence of the singularity inside, the big bang singularity lies *within* our cosmological horizon. An expanding universe is like a black hole turned inside out and upside down. The singularity in the beginning forms quite literally the past edge of our past light cone. So in principle it is there for us to see, writ large in the sky.

Of course we can't easily look all the way back to the beginning, for the constant scattering of light particles in the earliest stages of expansion obscures our view. Looking back at the big bang is a bit like looking at the sun. In the case of the sun, what we see as a fairly sharp contour is really the surface where the photons produced by the nuclear fusion reactions deep inside the sun scatter for the last time. From that surface, called the photosphere, the photons fly toward us, unhindered. But this photon scattering prevents us from seeing directly inside. The interior of the sun is opaque to particles of light, not transparent.

Likewise, the constant scattering of photons in the hot plasma that fills the early universe creates a fog preventing us from seeing all the way back to the beginning, at least with photon-gathering telescopes. The newborn universe only became transparent 380,000 years after the big bang, when it had cooled to a pleasant three thousand degrees Celsius. At this temperature it became energetically favorable for atomic nuclei to combine with electrons to form neutral atoms, leaving hardly any electrons for light particles to scatter against. As a consequence, photons began to travel unhindered through space, their wavelengths gradually stretching out a thousandfold, in sync with the expansion. What was red light at the outset reaches us today, billions of years later, as cold microwave radiation. Figure 2 in chapter 1 shows the sky map of that CMB radiation. This map gives us a snapshot of the universe from the moment in time that it became transparent. However, the microwave radiation also blocks our view of earlier epochs; the CMB map is the cosmological "inside-out" analogue of the sun's photosphere.

The singularity that bounds our past in general relativity highlights how deeply puzzling it is that the relic CMB radiation is distributed nearly uniformly throughout space. As I mentioned in chapter 1, the speckles in figure 2 represent temperature variations across the sky that are everywhere smaller than a ten-thousandth of a degree. Apparently the big bang played out in nearly exactly the same way in all regions of the observable universe. This is one of its curious biofriendly properties. In the case of the sun's photosphere, a nearly uniform temperature is just what one expects, since all the photons radiating from the sun's surface have been exchanging heat through interactions in its interior. Naturally, this leads them to acquire nearly the same temperature, just as cold milk rapidly attains a common temperature with hot tea (at least in the UK).

But it would appear that interactions couldn't have smoothed out the CMB radiation because there wouldn't have been enough time since the singularity for a physical process, even one moving at the speed of light, to level out any temperature differences before the ancient photons were liberated and began to fly freely through space. I illustrate this point in figure 19 with a slightly more accurate representation of an observer's past in a hot big bang universe compared to Ellis's sketch of it in figure 18. Microwave background photons reaching us from opposite directions in the sky start out at points A and B on our past light cone, but the past light cones of each of these points do not intersect back to the beginning. This means that no

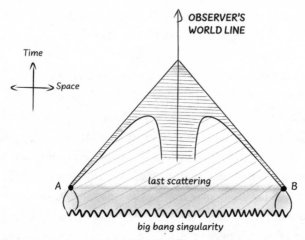

FIGURE 19. *Our past, according to the hot big bang model of the 1960s. We are located here and now at the tip of the cone. Microwave background photons reaching us from opposite directions in the sky originate from points A and B on our past light cone. These points are far outside each other's cosmological horizon: Their own pear-shaped past light cones don't overlap all the way back to the beginning. Yet we observe the temperature of the photons arriving from A and B to be the same to an accuracy of one-thousandth of 1 percent. How come?*

signal of light could have passed between A and B since the big bang. And since the speed of light sets the upper limit for how fast any signal can travel, this means that no physical process whatsoever could have established an interconnected environment encompassing A and B. Physicists say that the regions around A and B lie outside each other's cosmological horizon.

As a matter of fact, in the hot big bang universe of the 1960s, when we look at the CMB in directions separated by more than a few degrees in the sky, we are seeing patches of the universe that have yet to come into contact with each other. Our entire observable universe today would enclose no fewer than several million such independent cosmic horizon-size domains. This renders the near-perfect uniformity of the CMB radiation across the entire sky not just puzzling but outright mysterious. If Eddington or Einstein had known about it, this horizon riddle might well have confirmed their worst fears about the whole idea of a cosmic origin. It's as if the Norse Vikings had landed in North America only to find that the indigenous inhabitants spoke Old Norse.

This is a strange state of affairs. Hawking's singularity theorem predicts that the universe had a beginning, but it doesn't say how it began, let alone why it emerged from its explosive genesis with a nearly uniform CMB and so many other biophilic properties. Even more, it seems to place all questions about the ultimate origin of the universe and its design outside science, as if it is outsourcing these to Eddington's supernatural agency. It isn't necessary to philosophize about it—the theory of relativity predicts its own downfall. The big bang in Hawking's PhD is an event without explanation, because the singularity at the bottom of it signals the breakdown of time, space, and causality altogether. As the great Wheeler put it, "The existence of spacetime singularities represents an end to the principle of sufficient causation and so to the predictability gained by science."[5]

How can this be? How can physics lead to a violation of itself—to no physics? To untangle this we must take a closer look at what physicists actually mean when they say they predict what will happen.

EVER SINCE GALILEO and Newton, physics has been based on a dualism of some sort, in that it has relied on a fundamental separation between two distinct sources of information. First, there are the laws of evolution, mathematical equations that prescribe how physical systems change in time from one state to another. Second, there are boundary conditions, a concise description of the state of a system at a given moment in time. The laws of evolution take that state and evolve it, either backward or forward in time, to determine what the system was like at an earlier moment or what it will be at a later moment. It is the combination of the laws of evolution and the boundary conditions that yields the framework for prediction on which physics and cosmology pride themselves.

For example, imagine you want to predict where and when the next solar eclipse will occur. To do so we can apply Newton's laws of motion and gravity to describe the future trajectories of Earth and the moon. To put these laws to use, however, you must first specify the position and velocity of Earth and the moon relative to the sun (and Jupiter) at one particular moment in time. These data are boundary conditions. They describe the state of these two celestial bodies in the solar system at one specific moment. No one expects Newton's laws to explain why these positions are what they are at that moment. Instead, we measure what they are. With this information at hand, we then solve Newton's equations to determine their positions at future times, in order to predict when and where solar eclipses will occur, or at earlier times, to retrodict documented eclipses.

This example is representative of the way in which predictions in physics come about in general. Physicists assume that evolution is governed by universal laws of nature—laws that we seek to discover. But boundary conditions contain information specific to one or another system, so they aren't considered part of the laws. Boundary conditions in some sense serve to delineate the particular questions we ask of the physical laws. In effect, a given dynamic law like Newton's is constructed in such a way that it can accommodate a wide range of different boundary conditions. This gives the equations their universal character and the flexibility they need to account for a broad variety of phenomena. So the laws of physics are a bit like the rules of chess. However important those rules are, they tell you only so much about how any particular game will play out.

But is this separation between law-like dynamics and ad hoc boundary conditions a fundamental property of nature? This distinction is of course entirely natural and appropriate in laboratory situations where there is a sharp difference indeed between the experimental arrangement that one controls—the boundary conditions—and the laws one seeks to test by running the experiment. However, this distinction risks becoming a major embarrassment in cosmology, when we embed our experiments and experimenters, our planet, the stars, and the galaxies, in the much larger evolution of the universe as a whole. When we do this, boundary conditions on the original experiment are subsumed into the law-like evolution of the larger system, together with boundary conditions on the latter. Returning to the example of a solar eclipse, a holistic cosmologist would say that the velocities and positions of the planets at any given moment—the original boundary conditions—follow from their past histories, and that our planetary system itself is the outcome of the formation history of the solar system, which in turn arose from the condensation of remnants of prior stellar systems, whose seeds ultimately grew out of minute density variations in the primeval universe that came from . . . what?

When we arrive at the beginning we reach a paradox. What determines the ultimate boundary conditions at the origin of the universe? Clearly these are not up to us to choose, and we can't try out different conditions to see what kind of universes they produce. That is, the beginning of the universe poses a problem of boundary conditions that we do *not* control. Instead, very interestingly, the conditions at the big bang would appear to be dragged into the laws we seek to understand. Yet dualism in physics holds that boundary conditions aren't part of the physical laws. What's more, Stephen's singularity theorem that says spacetime and all known laws break down at the big bang would seem to confirm this view. Note that this paradox arises only in a cosmological context, for it is but when we conceive the evolution of the universe in its entirety that we have no earlier moment or bigger box available that we can use to specify boundary conditions.

More than any other physicist of his generation, Stephen felt that an understanding of the universe's beginning on the basis of science would require a genuine extension of the centuries-old framework for prediction in physics. Dynamics versus conditions, he felt, is too narrow a way of thinking about the world. Already in his PhD dissertation he put his finger on the problem, writing, "It is one of the weaknesses of Einstein's theory of relativity that although it furnishes dynamic field equations it does not provide boundary conditions for them. Thus Einstein's theory does not give a unique model for the universe. Clearly a theory that provided boundary conditions would be very attractive. . . . Hoyle's theory does just that. Unfortunately, its boundary condition excludes those universes that seem to correspond to the actual universe, namely the expanding models."

He elaborated on this point in his inaugural lecture as incumbent of the Lucasian Chair, nearly fifteen years later. The Lucasian Chair of Mathematics was established—with an allowance of 100 pounds a year—in 1663 by Henry Lucas, a former student of St John's College, philanthropist, and politician who sat in parliament for Cambridge University. From 1669 to 1702 it was held by none other than Isaac Newton (though Stephen often quipped it wasn't motorized at the time). Luckily for Newton, the deed that had established the Chair contained a provision that its holder should not be ordained in the Anglican Church. That meant that Newton, was exempt from swearing an oath on belief in the Trinity—something that would have been impossible for him.*

---

* Newton—nota bene as a Fellow of Trinity College—rejected the decision of the Council of Nicaea that the Father and the Son originated from one and the same substance.

Stephen was elected the seventeenth Lucasian Professor in 1979 and in his inaugural lecture, "Is the End in Sight for Theoretical Physics?" at the zenith of his confidence in the power of physical theory, he controversially predicted physicists would find the theory of everything by the end of the century. But he continued: "A complete theory includes besides a theory of dynamics also a set of boundary conditions." Elaborating on this point, he added, "Many people would claim that the role of science was confined to the first of these and that theoretical physics will have achieved its goal when we have obtained a set of local dynamic laws. They would regard the question of the boundary conditions of the universe as belonging to the realm of metaphysics or religion. But we shall not have a complete theory until we can do more than merely say that things are as they are because they were as they were."

So, Stephen, forever optimistic and ambitious, wasn't prepared to be taken hostage by his own singularity theorem. What the initial singularity is really telling us, he and others reasoned, is not that the big bang origin is bound to remain off-limits to scientific inquiry but that Einstein's description of gravity in terms of malleable spacetime fails under the extreme conditions that hold sway at the universe's birth. When we delve into the big bang, the small-scale randomness of quantum theory comes to the fore. One might say that space and time desperately want to break out of the highly constricted framework imposed by Einstein's deterministic theory. After all, for all its bending and warping, spacetime in general relativity remains an extremely constrained structure, consisting of a specific sequence of shapes of space, meticulously fitted together like Matryoshka dolls, one inside the other, to create four-dimensional spacetime.

More than anything else, Hawking's big bang singularity theorem demonstrated the gravity of the conflict between relativity and quantum theory. It reinforced Lemaître's intuition that cosmogenesis is a profoundly quantum phenomenon, and that if we are to stand a chance of untangling the riddle of design on the basis of science, we ought to somehow marry the principles of these two seemingly contradictory tales of nature. The crux of Stephen's vision was that this marriage would be much more than a mere refinement of the existing framework for prediction in physics, that it would require us to rethink that framework itself. His idea was to take physics beyond its age-old dualism of laws versus conditions, of evolution versus creation.

QUANTUM MECHANICS, THE second pillar of modern physics, has popped up several times by now. The theory has its roots in a number of mystifying experiments with atoms and light in the early years of the twentieth century that could not be explained by any stretch of the classical mechanics of Newton. Its formulation throughout the turbulent years of the early twentieth century stands as one of the finest examples of international collaboration in the history of mankind. Over the course of the century since, quantum mechanics went from one triumph to the next and became the most powerful and accurately tested scientific theory of all time. It applies to all known types of particles. From the fine details of how elementary particles interact to the fusion of atoms inside distant stars, the predictions of quantum mechanics perfectly match experimental data. And much as Maxwell's classical theory of electromagnetism laid the foundation for the second industrial revolution, the principles of quantum theory underpin today's technology. As a matter of fact, we may have seen only the tip of the iceberg of what quantum technology has to offer. In the near future physicists and engineers hope to exploit the intrinsic uncertainty of the microworld to store and process information in new ways, by manipulating individual quantum bits known as *qubits,* thus paving the way for the era of quantum computing.

The quantum revolution started in 1900 when the German physicist Max Planck suggested that any kind of body, when heated, emits radiation in discrete little packets he called *quanta.* Planck had been struggling to explain how much light of each color hot bodies radiate. He knew from Maxwell's classical theory that light is made of electromagnetic waves of different oscillation frequencies, corresponding to different colors. The trouble was that classical physics also predicted that the energy radiated by a hot body should be shared equally among waves of all frequencies. Since Maxwell's theory had electromagnetic waves of arbitrarily high frequencies, this meant that the total of radiated energy, summed over all frequencies, should be infinite, which was obviously impossible. This was

Lord Kelvin's second dark cloud hovering over classical physics. It became known as the "ultraviolet catastrophe" of classical physics, because the highest frequencies of visible light are violet, hence "ultraviolet" refers to very high frequencies.

In what he later described as "an act of despair," Planck proposed a bold new rule saying that light and all other electromagnetic waves can only be emitted in discrete quanta, and that the energy of each quantum was greater the higher the frequency of the waves. This sharply reduced the emission of waves of high frequencies, thereby avoiding an ultraviolet catastrophe. In 1905 Einstein went further and showed that electrons moving around in metals can only absorb light in discrete quanta, which he described as tiny particles, or photons. So, curiously, these early ideas on quanta implied that light had the properties of a wave but also of a particle, which caused quite some confusion.

The upheaval continued when, much as Planck had done for light, the Danish physicist Niels Bohr invoked quantization to explain the existence of stable atoms—another obvious property of the physical world. Bohr, who has the element bohrium named after him, trained at Manchester with the British physicist Ernest Rutherford, whose experiments had revealed that the inner structure of atoms consists mostly of empty space, with a tiny nucleus at its center. Rutherford thought of atoms as miniature planetary systems in which negatively charged electrons orbit a dense central nucleus carrying positive charge. Since unlike charges attract, the electrons are drawn into orbit around the nucleus. The trouble with this model was that, according to Maxwell's classical electromagnetism, orbiting electrons radiate energy, causing them to spiral inward and collide with the nucleus. This would mean that all atoms in the universe would rapidly collapse—and we wouldn't exist. To resolve this obvious contradiction with reality, Bohr suggested that electrons could not orbit at just any distance from the nucleus but only at certain specific, separated radii. That is, Bohr *quantized* the possible electron orbits. The resulting separation between allowed orbits made it impossible for electrons to give in to their penchant to spiral inward, saving atoms from rapid—theoretical—collapse, a discovery for which he received the Nobel Prize in 1922.

The quantum pioneers gathered in Brussels in 1911, at the invitation of the Belgian industrialist Ernest Solvay, for one of the very first international physics conferences. This was a time when internationalism was cultivated as a kind of domestic policy in Belgium. Solvay was a liberal visionary who had made his fortune by converting his invention of a new process to synthesize sodium carbonate into a far-flung industrial network. After retiring from business he became a fervent alpinist, who climbed the Matterhorn on several occasions and also got King Albert I of Belgium interested in climbing, though eventually with unforeseen and disastrous consequences.

The First Solvay Council was to achieve mythical status, as it was here, in the plush Hotel Métropole in central Brussels, that scientists finally grasped the paradigm-shattering implications of the early quantum ideas. The council, presided over by the eminent Dutch physicist Hendrik Lorentz, defined the watershed between the classical physics of the nineteenth century and the quantum physics that would come to dominate the twentieth. Lorentz's opening address reverberates with the anguish that this master of classical physics felt at the first glimpses of the quantum world. "Modern research has encountered more and more serious difficulties when attempting to represent the movement of smaller particles of matter. . . . At the moment, we are far from being completely satisfied. . . . Instead, we now feel that we reached an impasse; the old theories have been shown to be powerless to pierce the darkness surrounding us on all sides."[6] But while the first Solvay meeting aired everything it settled nothing. The participants remained confused and divided over whether classical physics could somehow be patched up to accommodate quanta. Einstein captured the mood, saying, "The quantum-disease looks increasingly hopeless. Nobody really knows anything. The whole affair would have been a delight to Jesuit fathers. The council gave the impression of a lamentation at the ruins of Jerusalem."

ALL THIS CHANGED in the mid-1920s when a new generation of quantum physicists recast the mechanics of atoms and subatomic particles in a fundamentally new form—quantum mechanics.

A central tenet of the new mechanics was the famous uncertainty principle of the German prodigy Werner Heisenberg: You cannot know both the exact position and the velocity of a particle at the same time. As he put it, "The more precisely the position [of a particle] is determined, the less precisely the momentum [or velocity] is known in this instant, and vice versa."[7] The very best one can hope for in quantum mechanics is a fuzzy view where you know the positions and velocities of particles approximately.

As a matter of fact, all measurable quantities are subject to quantum uncertainty, to an extent described by Heisenberg-like principles. Such quantum uncertainty cannot be reduced by looking harder or by measuring the properties of particles in an ingenious way that evades Heisenberg's principle. In this respect it differs from, say, random movements in the stock market, which only appear unpredictable because humans do not have all the information they need to work out how stocks will behave. Heisenberg's quantum uncertainty, by contrast, is held to be fundamental. It sets stringent limits on the amount of information that can be extracted from physical systems, even in principle. So, intriguingly, quantum mechanics appears to be a theory about what we know but also about what we can't know. This strangeness will prove to be an all-important property when we will be considering the multiverse from a quantum perspective in chapters 6 and 7.

The awesome achievement of quantum physicists in the mid-1920s was to integrate this quantum fuzziness into a proper mathematical formalism. Not surprisingly, the resulting theory portrayed a far more slippery and fluid view of mechanics than what our classical grasp of it suggested. Quantum mechanics, for example, abandoned the old dream of scientific determinism, the idea that science should be able to make definite and precise predictions about the future course of events. The theory replaced that notion with the idea that we can only predict probabilities for different possible outcomes of measurements. Quantum mechanics holds that if one runs the exact same experiment over and over, one will, in general, not get the same results.

Rutherford may well have been the first to glimpse this fundamental layer of indeterminism woven into the microworld. In the year 1899, to study the inner structure of atoms, Rutherford bombarded a thin gold foil with alpha particles emitted from a radioactive source like uranium. As he observed the flashes of light, Rutherford soon recognized that the direction and arrival times of the alpha particles were random. According to quantum mechanics, this is because while uranium nuclei have a definite, calculable *probability* of decaying during a fixed time interval, it is impossible to know in advance when a *given* nucleus will decay. Quantum mechanics predicts the likelihoods of different arrival times and trajectories of alpha particles emitted in the decay of a radioactive sample, but it also says there is nothing that we know—or can hope to know—that would allow us to predict where and when a particular alpha particle will go. The strength of the theory—and its strangeness—is that it engraves this irreducible kernel of uncertainty and randomness suffusing the microworld into its basic mathematical underpinnings. The laws of quantum mechanics yield betting odds, really, rather than predictions for definite outcomes of observations. They force us to accept that the best one can do is predict probabilities for various results.

This key feature of the theory is perhaps clearest in the formulation devised by the Austrian physicist Erwin Schrödinger. In 1925 Schrödinger wrote down a fascinating equation that describes particles not as minuscule pointlike objects but as extended wavelike entities. But, and this is crucial, the waves the Schrödinger equation speaks of aren't physical waves. Schrödinger did not say that particles are somehow smeared out over space. The waves of quantum mechanics are a bit more abstract, they are more like "waves of probability" that describe different *possible* positions of pointlike particles. The way Schrödinger's formalism accounts for quantum uncertainty is that locations where the wave's values are large are locations where the particle is likely to be found. Locations where the wave's values are small are locations where the particle is unlikely to be found. One might say that quantum waves are a bit like crime waves: Just as the arrival of a crime wave in your town means you are more likely to find that a crime will have been committed, so does an electron wave that peaks in your apparatus mean you are likely to detect an electron.[8]

Given a particle's wavelike profile at a certain moment—its *wave function,* as physicists say—the Schrödinger equation predicts how it will evolve in the course of time, going up in some locations and down elsewhere. Hence quantum theory adheres to the dualistic scheme of prediction that I outlined above, with law-like dynamics versus boundary conditions. The Schrödinger equation is a law of evolution and this needs a condition in the form of the particle's wave function at a given moment to tell it what it is that evolves. The key difference with the classical mechanics of Newton and Einstein is that the laws of quantum theory predict only probabilities for how things will be at a later moment but no certainties. The dual nature of the basic framework for prediction, however, remains unchanged, as if cast in stone.

Now, as they are waves of probability, we can glean wave functions only indirectly. Schrödinger's quantum waves describe the world at some kind of preexistence level. Before one measures a particle's position there is no sense in even asking where it is. It does not have a definite position, only potential positions described by a probability wave that encodes the likelihood that the particle, if it were examined, would be found here or there. It is as if we compel particles to assume a position by looking at them, that there is a tangible physical reality only to the extent that we interact with the world by observing and experimenting. "No question, no answer!" is how Wheeler once put it.

The celebrated double-slit experiment provides a vivid illustration of this nebulous, wavy nature of the quantum world. Its setup, shown in figure 20, consists of a gun firing electrons at a barrier containing two narrow parallel slits and a screen placed behind the barrier on which the electron impact locations are recorded by a tiny flash. Suppose one tunes the gun so that it fires only one electron at a time, say, every few seconds. Then one finds that every electron making it through the barrier arrives at a particular spot on the screen, creating a little flash. So the individual electrons do not spread out. This is the particle nature of the electron—no surprises so far. However, if one lets the experiment run for a while, registering the impact locations of many electrons, then gradually an interference pattern builds up on the screen, consisting of a series of bright and dark stripes, that reminds us of intermingling wave fragments (see figure 20). Similar interference fringes have been observed in double-slit experiments with other elementary particles, light particles, atoms, and even molecules.

These interference patterns indicate there's something profoundly wavelike associated with individual particles that knows about both slits. It is that something that a particle's wave function captures. By describing electrons not as traveling particles but as propagating waves of probability, the Schrödinger equation predicts that, much like interfering waves on a lake, fragments of an electron's wave function coming out of different slits will intermingle, yielding a pattern of high and low probabilities for where on

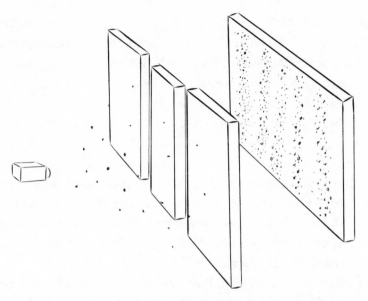

FIGURE 20. *The famous double-slit experiment, first carried out with electrons in 1927 at Bell Labs, demonstrated that electron particles have wavelike properties. Quantum mechanics explains the interference pattern on the screen on the right by describing each individual electron as a propagating wave function that splits at the slits in the middle, spreads, and intermingles with itself on the far side, creating a pattern of high and low values corresponding to high and low probabilities for where on the screen it will land.*

the screen each individual electron will land. Where the wave fragments emerging from both slits arrive in step with each other, they will reinforce; where they arrive out of step, they will cancel. When particle after particle is fired, their cumulative landing positions conform to the probabilistic profile encoded in each individual particle's wave function, thereby building up the observed interference pattern. It is at the deeper level of its probability wave, therefore, that each individual particle senses both slits.

THE PROBABILISTIC PREDICTIONS of quantum theory agree with every particle experiment that has ever been carried out. But its rules violate common sense. The quantum description of particles as abstract wavelike superpositions of mutually contradictory realities doesn't match up with our daily experience that objects are either in one place or another. Of course this (sometimes) bothered the founding fathers of quantum theory. In the words of Erwin Schrödinger, the quantum universe is "not even thinkable," for "however we think it, it is wrong; not perhaps quite as meaningless as a triangular circle, but much more so than a winged lion."[9]

Two decades later, this counterintuitive nature of quantum mechanics also bothered Richard Feynman. A student of the visionary Wheeler, Feynman became one of the twentieth century's most influential physicists, making major contributions from particle physics to gravity and computational science. Feynman acquired worldwide fame on the presidential Rogers Commission investigating the *Challenger* disaster when, during a televised hearing, he demonstrated the failure of the shuttle's O-rings. Afterward he pointedly warned in the commission's report, "For a successful technology, reality must take precedence over public relations, for nature cannot be fooled."

If Wheeler was the dreamer, then Feynman was the doer. Wheeler looked to the far past and the distant future, to the foundations of physical reality and the fundamental nature of scientific inquiry. Feynman strove to make physics work here and now, preaching that all he was interested in was trying to find a set of rules that supplied predictions that one could verify experimentally, and not go very far beyond that.[10] In this spirit, in the late 1940s, Feynman set out to develop a more intuitive and practical way of thinking about quantum particles and their wave functions. His idea was to imagine that particles are sort of localized objects but that they follow all possible paths when they move from one point to another (see figure 21). Classical mechanics supposes that objects take a single path

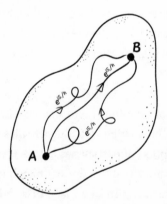

FIGURE 21. *Newton's classical mechanics dictates that particles follow a single path between two points, A and B, in spacetime. Quantum mechanics holds that a particle takes all possible paths. Quantum theory predicts that there is only a probability for arriving at B, which is a weighted average over all ways to get there.*

FIGURE 22. *Richard Feynman (right) talking
with Paul Dirac at the 1962 Relativity Conference held in Warsaw, Poland.*

through spacetime. Consequently, a classical system has a unique and
well-defined history. Quantum mechanics, Feynman argued, takes a more
expansive view of history and asserts that all possible pathways play out
simultaneously, although some would be much more likely than others.

For example, Feynman's understanding of the double-slit experiment
was that individual electrons follow not one but every possible path from
the gun to the screen. One path takes the electron through the left slit,
another through the right, and yet another trajectory might take it first
through the right, back out through the left, into a U-turn, and through
the left slit once more. Every single possible path—aka history—of the
electron, no matter how absurd, must be considered, Feynman advanced,
and all those pathways contribute to what we see on the screen. Feynman's
description of the electron's motion resembles a bit the alternative route
suggestions in a GPS device, except for the highly unusual—and profoundly quantum—phenomenon that, unlike most taxi rides, electrons
take all routes, and this is how quantum uncertainty enters in his scheme.
As Feynman put it, "The electron does anything it likes. It just goes in any

direction at any speed, forward or backward in time, however it likes, and then you add up the amplitudes [of their paths] and it gives you the wave function."[11]

To predict the probability that an electron arrives at a given point on the screen, Feynman tagged a complex number onto every path that specified its contribution to the probability and also how it interfered with paths nearby. This number basically endowed each individual path with the mathematical properties of a wave fragment. Next he wrote down a beautiful equation, an alternative to Schrödinger's, that constructs a particle's wave function by adding all paths ending up at every point. The telltale interference pattern on the screen comes about from the intermingling of the trajectories in Feynman's sum that are emerging from both slits. Mathematically, this is because the complex number assigned to every pathway means that different paths can enhance or diminish each other, just as wave fragments do.

Feynman's description of the two-slit situation exemplifies that there is no hope to find out from observations on the screen alone through which slit the electron actually came. This should not come as a surprise. By having not one but many histories playing out, quantum mechanics obviously limits what we can say about the past. The quantum past is inherently fuzzy. It isn't the kind of sharp and definite history we usually think of when we consider the past.[12]

Remarkably, Feynman's *sum-over-histories* scheme yields a perfectly viable and precise way of thinking about quantum theory in general. Fittingly it became known as the *many histories* formulation of quantum theory. In Feynman's view, the world is a bit like a medieval Flemish tapestry—a woven texture of crisscrossing paths that stitch a coherent picture of reality from the threads of a myriad of possibilities.

STEPHEN WAS FULL of admiration, both for Feynman himself and for his sum-over-histories approach to quantum physics. The two saw much of each other in the 1970s during Stephen's regular visits to Caltech. *Quite a character,* he once told me, *but a brilliant physicist.*

Feynman's framework proved a crucial stepping-stone for physicists to begin to think about quantum mechanics outside of its original home of the subatomic world. His approach showed that despite appearances, there need not be a fundamental contradiction between classical and quantum mechanics. The reason is that the sum-over-histories scheme applies both to small and to large objects, but that for larger objects, the only trajectories with any significant probability are those lying everywhere along the single path predicted by Newton's classical laws of motion. So there isn't a fundamental dichotomy between the micro- and macroworlds after all. It is just that for macroscopic objects, the microscopic jittering averages out to something definite and deterministic, and that something is the path of classical motion. Classical determinism, that is, *emerges* from the collective behavior of random microscopic quantum histories. By contrast, delve into the microscopic realm and more and more of the random crisscrossing becomes relevant.

All of these insights—and the astonishing successes of quantum theory—meant that the classical worldview faded away. Many physicists began to believe that quantum theory, which began as a theory of subatomic particles, applies to all objects on all scales. In the 1960s, Wheeler and his gang began to envision even spacetime as a quantum foam, writhing with bubbling baby universes and with wormholes popping up and dissipating again, somehow averaging out on macroscopic scales to the definite fabric of the classical theory of general relativity.

Stephen too ventured with Feynman's sum-over-histories framework into the realm of gravity. He had been introduced to Feynman's scheme by Jim Hartle, who had learned it from Feynman himself as a graduate student in Caltech, the Marine Corps of graduate schools at the time. Jim attended Feynman's classes while assisting him with his lecture demonstrations— among them the celebrated Bowling Ball Demonstration—and with the redaction of *The Feynman Lectures on Physics,* the most famous physics textbook ever, brilliant in its comprehensive exposition—though rarely turned to.

In 1976, Jim and Stephen were able to describe Hawking radiation from black holes as particles leaking out of the horizon by summing up, à la Feynman, all possible paths that particles could take to escape from a black hole.[13] Encouraged by this result, they turned their attention to the more challenging and confusing big bang singularity—the cosmic analogue of point A in figure 21. For a particle, quantum uncertainty means that its position and velocity are somewhat imprecise. Applied to space-time, then, quantum uncertainty should mean that space and time themselves are somewhat fuzzy, with quantum jitters smearing out points in space and moments in time. In nearly all of the observable universe such spacetime fuzziness would be extremely limited and hence utterly irrelevant, but back in the earliest stages of the universe, with matter densities and spacetime curvature rising without bounds, quantum uncertainty would appear to have been hugely important. Reasoning along these lines, Stephen imagined that in the very early universe, quantum effects would have blurred the very distinction between space and time, causing them to suffer a bit of an identity crisis, with intervals of time sometimes behaving like intervals of space and vice versa. What's more, Jim and Stephen boldly proposed that the Feynman sum over all this frenzied spacetime fuzziness could be done and that the resulting wave function could be expressed in an elegant geometric manner.

To get a feel for their *wave function of the universe,* look at figure 23. This is the same schematic representation of an expanding universe that I showed in figure 14 in chapter 2, but this time around I have run the movie of this universe backward in time. Figure 23(a) reminds us of what

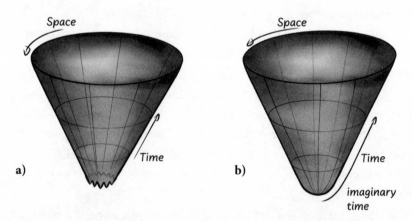

FIGURE 23. *Classical and quantum evolution of an expanding universe shown here as a one-dimensional circle. Panel (a): In Einstein's classical theory of gravity, the universe originates in a singularity, at the bottom, at which the curvature is infinite and the laws of physics break down. Panel (b): In Hartle and Hawking's quantum theory, the singular origin is replaced with a smooth and rounded bowl-like shape, complying with the laws of physics everywhere.*

happens if we blindly trusted Einstein's classical relativity theory: Space is getting smaller in the past and at some point dissolves into a singular state of infinite warping and curvature, dragging time in its demise.

But Jim and Stephen argued that that is not what actually happens. According to them, quantum mechanical effects dramatically alter the evolution when we run the clock that far backward. In fact, they envisaged that the blurring of space and time would effectively rotate the vertical time direction into an additional horizontal space direction. Now, this opened up a whole new possibility for the origin of the universe: These two dimensions of space could combine to form a smooth two-dimensional spherical surface, somewhat like the surface of Earth. Figure 23(b) depicts this quantum evolution. We see that the singularity at the bottom of the classical universe, that event without a cause that seemingly put the beginning outside science, is replaced here by a smooth and rounded quantum origin complying with the laws of physics everywhere.

This was a supremely original idea. The crux of Jim and Stephen's proposal was that an expanding universe has no singularity in the past because the dimension of time dissolves into quantum fuzziness on our way back toward the beginning. At the base of the bowl in figure 23(b), time has become space. Hence, the question of what might have come before has become meaningless. "Asking what came before the big bang would be like asking what lies south of the South Pole," Hawking summed up the theory, and he went on to refer to their quantum cosmogenesis as the *no-boundary* proposal.[14]

Stephen's no-boundary hypothesis blends together two seemingly contradictory properties. On the one hand, the past would be finite—time does not extend backward indefinitely. On the other hand, there would be no beginning either, no first moment when time is somehow switched on. If you were an ant crawling along the surface in figure 23(b), looking for the origin of the universe, you wouldn't find it. The spherical base of the bowl represents the past limit of time, but it doesn't mark an instant of creation. Any attempt to pin down an actual beginning in the no-boundary theory is futile—lost in quantum uncertainty.

From an aesthetic point of view there's something attractive about the way the no-boundary hypothesis meanders around the conundrum of the zero of time. The bowl at the bottom of spacetime in the theory has the feel of a geometric version of Lemaître's primeval atom. Taking his cue from Hamlet, who said, "I could be bounded in a nutshell and yet count myself a king of infinite space," Hawking saw the newborn universe as a nutshell in his hand.

WHEN JIM AND Stephen submitted their manuscript "The Wave Function of the Universe" for publication in the *Physical Review* in July 1983, things did not go smoothly. The first referee recommended against publication on the grounds that the authors employed an outrageous extrapolation of Feynman's sum-over-histories formulation of quantum theory to the entire universe. Jim and Stephen then requested a second opinion. The second referee said he agreed with the first one that the extrapolation envisioned by the authors was outrageous indeed. Nevertheless, he continued, the manuscript should be published, "because this will be a seminal paper."[15] And that is exactly what happened. Five decades after Lemaître's 1931 manifesto in which he called for a quantum perspective on the origin of time, Jim and Stephen's landmark discovery turned Lemaître's daring vision into a genuine scientific hypothesis. Their universal wave function sparked a wave of interest in the quantum foundations of cosmological theory that would become a key to physicists' quests to unravel the riddle of design.

As a matter of fact, the no-boundary hypothesis emerged out of a whole new approach to studying the quantum nature of gravity that Stephen, working with his first generation of students, had been developing throughout the 1970s. The Cambridge approach to quantum gravity was based on Einstein's geometric language but, strikingly, made use of curved shapes of four space dimensions, without a time direction, rather than the warped spacetimes of relativity.

In Einstein's classical relativity theory, space is space and time is time. To be sure, space and time are unified in four-dimensional spacetime, as the diagrams that I have shown vividly demonstrate, from Minkowski's empty spacetime to Penrose's black hole geometry. But in all these diagrams it is easy to tell the difference between space and time: The arrow of time points everywhere within the future light cone, whereas the space directions do not (see, e.g., figure 8). Now Stephen had a vision that warped geometries with four space dimensions encapsulated profound quantum properties of gravity. Consequently, his research program became known as the *Euclidean approach to quantum gravity*, after the ancient Greek mathematician Euclid, who was the first to systematically study the geometry of spatial dimensions.

Geometrically, the transformation of time into space amounts to rotating the time direction by 90 degrees. This is evident from the quantum panel in figure 23 where "early on," down at the bowl, time starts "flowing" in the horizontal plane, on equal footing with the circular dimension of space. This time-into-space rotation is often described as making time imaginary, because mathematically the rotation corresponds to multiplying time by an imaginary number, namely the square root of minus one. Obviously this operation renders void all notion of normal evolution. It would be no use whatsoever to set your alarm to $7\sqrt{-1}$ A.M. to catch your morning train. Even a process as slow as Brexit played out in real time. "Any subjective concept of time related to consciousness or the ability to perform measurements would come to an end," Stephen proclaimed. But by bending Einstein's curved geometries further than anyone had ever done, from real time to imaginary time, he had identified a thrilling new inroad into the quantum realm of gravity.

Take a black hole. Penrose's drawing of a black hole in figure 11 in chapter 2 depicts the geometry of a classical black hole, one that exists in real time. The geometry of a quantum black hole in imaginary time has a very different shape. It is more like the surface of a cigar depicted in figure 24. Moving "forward" in imaginary time in this black hole geometry corresponds to going around the circle. The tip of the cigar represents the horizon of the black hole. Nothing lies beyond it, to the left of it in figure 24, so unlike a black hole in real time, its Euclidean counterpart possesses no singularity where the theory breaks down. Much like the no-boundary proposal replaces the singular beginning of a classical universe with a rounded quantum origin, the Euclidean description of a black hole has a smooth and gentle geometry that complies with the (quantum!) laws of physics everywhere. Working with Euclidean shapes of black holes, Stephen and his Cambridge group were able to understand the deep reasons why black holes aren't completely black but radiate quantum particles much like ordinary bodies with a certain temperature.[16]

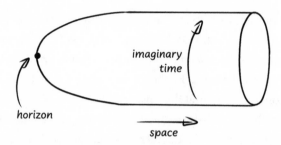

FIGURE 24. *Black holes have the shape of a cigar when one considers them in imaginary time. The horizon of the black hole corresponds to the tip of the cigar on the left. Geometric smoothness of the tip ties in with the size of the circular, imaginary-time dimension on the right. The latter, in turn, determines the temperature of the black hole and hence the intensity of the Hawking radiation that escapes in real time.*

The power of Euclidean geometries to describe quantum properties of gravity left a very strong impression on Stephen. His imaginary-time method became the cornerstone of his efforts to marry the principles of gravity with those of quantum theory in order to unlock the secrets of the big bang. "One could take the attitude that quantum gravity and indeed the whole of physics is really defined in imaginary time," he declared at one point. "It is simply a consequence of our perception that we interpret the universe in real time."[17]

In ordinary quantum mechanics, without gravity, rotating time into space is a standard trick that physicists use to carry out Feynman sums over particle histories. This is because adding paths in imaginary time simplifies the complicated Feynman sum. At the end of the calculation, physicists then rotate one of the space dimensions back into real time and read off the resulting probabilities for the particle to do this or that. But Jim and Stephen didn't want to rotate back to real time. The audacity of their no-boundary proposal was that when it comes to the origin of the universe, the transformation of time into space isn't merely a clever calculational trick but fundamental. The story of the universe, the theory holds, is that once upon a time there was no time.

That being said, there is something Einsteinian about the no-boundary idea. In 1917, when Einstein was pioneering relativistic cosmology, he was at a loss about the boundary conditions at the spatial edge of the universe. Einstein concluded that it would be much easier if space had no boundary. Thus he was led to conceive of our spatial universe as a giant three-dimensional hypersphere, which, much like the two-dimensional surface of a normal sphere, has no edge or boundary. With their no-boundary hypothesis, Stephen and Jim sorted out the problem of the boundary conditions at the zero of time in a similar Einsteinian fashion, by eliminating the initial boundary altogether.

It is remarkable that Stephen developed his geometric approach to quantum gravity during the period when he was losing the use of his hands for writing equations. This loss may well have encouraged him in his attempt to cast the unfathomable quantum realm of gravity in the language of geometry and topology, which he could visualize on the blackboard and, to some extent, manipulate in his head. Visualization was central to Stephen's thinking indeed. Working with Stephen meant working with shapes and pictures to represent the physical essence of mathematical relationships. Very early on in our collaboration, I got a taste of the way he handled calculations without the ability to write equations when I went to see him one time in the hospital where he was recovering from a lifesaving operation. We talked for a bit about the ordeal he had just gone through, but then Stephen requested that I go and find him a whiteboard in the hospital wing. When I finally got one, he asked me to draw a circle. By the end of the afternoon the circle represented the edge

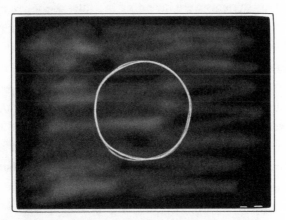

FIGURE 25. *The evolution of an expanding universe, in imaginary time.*

of the disk that you get when you project the expanding quantum evolution in figure 23(b) down onto a flat plane. The origin of the universe lies at the center of the disk, and the universe today corresponds to the circle. All this in imaginary time, of course.

Stephen elaborated his Euclidean approach to quantum gravity to the point that it gave him insights that would have been nearly impossible to obtain in any other way. The no-boundary hypothesis is perhaps the most striking example of this. But the time-into-space rotation at the heart of it also meant that it was very difficult to grasp what was really going on at the beginning of the universe. The bowl at the bottom of spacetime says we ought to let go of our cherished idea that there has always been a time around to give meaning to before and after. But it says frustratingly little about what—if anything—might be happening in the absence of time, or what kind of microscopic quantum fuzz, when added up, produces the bowl-like geometry in the first place. It is as if the theory is trying to tell us that we shouldn't ask such difficult questions.

So physicists complained that Stephen's creative use of Euclidean geometries was like magic. His entire approach was often dismissed as a Cambridge eccentricity. Why should time behave in such strange ways? Part of the problem was that the Euclidean framework isn't a fully fledged quantum theory of gravity but a semiclassical amalgam of classical and quantum elements put together without clear mathematical guidelines. Stephen and his students were inventing the rules as they went along. As Harvard theorist Sidney Coleman phrased it, after an attempt of his to argue on the basis of the Euclidean approach that the cosmological constant should be zero: "The Euclidean formulation of gravity is not a subject with firm foundations and clear rules of procedure; indeed it is more like a trackless swamp. I think I have threaded my way through it safely, but it is always possible that unknown to myself I am up to my neck in quicksand and sinking fast."[18] Stephen, however, remained undeterred. "I'd rather be right than rigorous," he countered. He just had this very strong intuition that Euclidean geometries provided a uniquely powerful gateway into the most extreme patches of the universe—into black holes and the big bang. Today, nearly forty years on from his pioneering work on quantum cosmology, the no-boundary hypothesis continues to generate great interest, profound confusion, and heated controversy, but with, as yet, no viable alternative description for our deepest origins in sight.

IN WHAT SEEMS to have been an intended resonance, Stephen first proposed that the universe had no boundary, no definite moment of creation, at a meeting of the Pontifical Academy of Sciences in the Vatican in October 1981. The academy's stated goal is to advise the Vatican on scientific matters and to enhance the mutual understanding between science and religion. To this end the Pontifical Academy had invited scientists from all over the world to the picturesque Casina Pius IV, its seat in the midst of the botanical gardens behind St. Peter's Basilica, for a weeklong debate on the topic "Cosmology and Fundamental Physics."[19] But the big bang proved a delicate point. Early on in the week Pope John Paul II told the assembled scientists: "Every scientific hypothesis on the origin of the world, such as that of the primeval atom giving rise to the entire physical universe, leaves open the problem of the beginning of the universe. Science by itself cannot resolve such a question. It requires knowledge beyond physics and astrophysics, which is known as metaphysics. Above all, it requires knowledge which comes from the revelation of God."[20] As if he were responding to the pope's allocution, Stephen, in a stunning lecture, "The Boundary Conditions of the Universe," put forward the bold idea that there might not have been a beginning. "There ought to be something very special about the boundary conditions of the universe and what can be more special than the condition that there is no boundary," he propounded, leaving his audience dumbfounded.

The no-boundary wave function of the universe that grew out of this was—and indeed still is—a physical law of a radical new kind. It is neither a law of dynamics nor a boundary condition but a blend of both, embodying a new sort of physics altogether. Earlier on I mentioned that classical physics and the ordinary quantum mechanics of particles alike subscribe to the orthodox dualistic framework of prediction that separates laws from initial conditions. Not so for no-boundary cosmology, which abandons this dichotomy in favor of a more general framework that treats initial conditions and dynamics on equal footing. According to the no-boundary hypothesis, the universe doesn't quite have a point A where external conditions would have to be specified.

As a matter of fact, something like this had been long in the making. In his 1939 Edinburgh lecture, Paul Dirac already anticipated the demise of dualism in physics. "The separation [between laws and conditions] is so unsatisfactory philosophically, as it goes against all ideas of the unity of Nature, that I think it is safe to predict it will disappear in the future, in spite of the startling changes in our ordinary ideas to which we should then be led." Four decades later, the no-boundary proposal did precisely this.

With their hypothesis, Jim and Stephen achieved what so many great thinkers, from Kant to Einstein, deemed impossible. Bridging the age-old chasm between evolution and creation, the theory put the question of the origin of the universe securely within the natural sciences at last. It offered a chance to resolve the conundrum of the universe's beginning conclusively. This was obviously very appealing. Stephen really felt he had found a way around the singularity—that he had cracked the great enigma of existence.

Unlike Lemaître, he did not refrain from implicating theology into his cosmogony. "The universe would be completely self-contained and not affected by anything outside itself," he wrote in *A Brief History of Time*. "It would neither be created nor destroyed. It would just be. . . . What place then for a creator?" The no-boundary theory, Stephen argued, dispensed with the need for a *primum movens* that set the universe going, because it showed that the universe could have been created from nothing. Of course, the God of the gaps Stephen evokes in this passage in *A Brief History of Time* is a far cry from Lemaître's *Deus Absconditus,* the Hidden God, hidden even in the beginning of creation.

To be clear, this was the early Hawking talking, who adhered to the metaphysical position of Einstein. Like Einstein, the early Hawking presumed that the mathematical laws of physics had some sort of existence that superseded the physical reality they governed. Indeed Einstein had hated the idea of a big bang in large part because it appeared to undermine this ideal. Whereas Stephen's singularity theorem had appeared to confirm Einstein's suspicions, the no-boundary bowl that replaced the singularity in quantum cosmology seemed to let one get a grip on the beginning and all the while uphold Einstein's idealism. This was an exciting prospect indeed.

However, much as Einstein's own theory of relativity took him by surprise, the no-boundary hypothesis would come to surprise Stephen. This early version of the no-boundary proposal was in no way radical enough!

FIGURE 26. *Stephen Hawking with some of his academic offspring at his sixtieth birthday party in King's College, Cambridge.*

CHAPTER 4

ASHES AND SMOKE

Creía en infinitas series de tiempos, en una red creciente y vertiginosa de tiempos divergentes, convergentes y paralelos . . . ; en algunos existe usted y no yo; en otros, yo, no usted . . . ; en otro, yo digo estas mismas palabras, pero soy un error, un fantasma.

He believed in an infinite series of times, in a growing, dizzying net of divergent, convergent and parallel times. . . . In some you exist, and not I; in others I, and not you. . . . In still another, I utter these same words, but I am a mistake, a ghost.

—JORGE LUIS BORGES, *El jardín de senderos que se bifurcan*

STEPHEN'S CAMBRIDGE SCHOOL OF RELATIVITY WAS LIKE A ROCK group: informal, out of touch with everyday reality, and radical in its ambition to change the world.

Its home base at DAMTP, the Department of Applied Mathematics and Theoretical Physics, had been founded in 1959, by the applied mathematician George Batchelor. At first DAMTP was housed in a wing of the Cavendish Laboratory, the famous lab where, in 1897, J. J. Thomson discovered the electron and where, in 1953, Watson and Crick deciphered the helical structure of DNA.* Then, in 1964, DAMTP moved to the Old Press Site opposite the Fitzbillies bakery, in between Silver Street and Mill Lane, and this is where I first met Stephen. The Victorian building, unassuming from the outside, had the most illogical floor plan inside, with a labyrinth

---

* Cambridge folklore has it that Watson and Crick actually discovered DNA in the Eagle, a pub across the street.

of dimly lit corridors leading to lecture rooms, dead ends, and dusty offices. We loved the place.

DAMTP's beating heart was its "common area." With high ceilings supported by pillars, stern portraits of former Lucasian Professors gazing down from the walls, vinyl armchairs, and a bulletin board stacked with posters announcing student parties or scientific conferences, this is where Dennis Sciama in the mid-1960s introduced an almost compulsory daily teatime ritual. At four P.M. sharp the lights would go on, cups would be lined up on the counter like a toy army, and tea would be served. In an instant the hall would be bustling with activity. Theoretical physics, after all, is a profoundly social pursuit.

Stephen would emerge from behind the olive-green door of his office with his right hand holding his clicker and his left hand wrapped around his steering knot, navigating his wheelchair through the crowd— occasionally rolling over someone's toes—to join in the conversation. Discussions took place around low tables with washable white tops, ideally suited for scribbling equations and trying out new ideas on one another. The tea itself was pretty bad, but the occasion enabled excellent science by bringing people together. Robert Oppenheimer, of atom bomb notoriety and former director of the Institute for Advanced Study at Princeton, once said, "Tea is where we explain to each other what we don't understand." For years DAMTP's tea served precisely this purpose, turning its common area into an international hub for the latest on theoretical physics.

My own daily teatime ritual with Stephen forged a bond that ran far deeper than the typical teacher-student relationship. Often our discussions would continue long after the common room had emptied again, carrying over into the evening either at the Mill, DAMTP's after-work pub hangout on the river Cam, or over dinner at his house in Wordsworth Grove.* Working with Stephen was an all-in package. There wasn't much of a separation between his professional and personal life. In many respects, he treated his circle of close collaborators as a second family.

Now, John Wheeler once said there are three ways to do great science:

---

* In those days he usually served very hot curry.

the ways of the mole, the mutt, and the mapmaker. The mole starts at one point in the ground and systematically goes forward. The mutt sniffs around and is led on from one clue to another. The mapmaker, finally, conceives of the overall picture, has an intuition for how things fit together, and so finds his way to where new understanding lies. Hawking, in my opinion, was a mapmaker.

Whereas Sciama had been supremely effective in connecting people around key open problems in theoretical physics, Stephen derived from his map a clear agenda of his own. But he relied on us to fill in the blank spaces on his map. From day one Stephen expected us to work with him to transform the grand intuitive picture in his head into fully fledged research projects and carry them out. As a result he kept us much closer than most advisers do.

Obviously, since communication via his speech synthesizer was necessarily limited—not only in terms of words but especially when it came to manipulating equations—Stephen could not provide much guidance throughout the detailed calculations involved. Instead he laid out general directions and adjusted these as we went along. That said, it could be frustratingly challenging to navigate Stephen's map with only his brief, seemingly encrypted directives in hand, but that in itself was stimulating because it forced us, his students, to think creatively and independently.

And he trusted us. Stephen radiated an irrepressible confidence that we could solve these difficult cosmic riddles. The same steely determination that allowed him to persevere, despite his debilitating disease, manifested itself as a certain stubbornness in his scientific work. Whenever I was in the depths of despair, when a line of research had crashed and I felt I had almost proved that what we were trying to do was impossible, Stephen would come along and unfold his mental map to offer a new perspective, pulling us out of the pit of doom and onto a fresh new track. That was Hawking's modus operandi—to reach for the deepest problems, keep attacking them from different angles, and find a way forward.

He throve in his role as a deus ex machina adviser. Moreover, his trust and quick wit and the warmth he exuded meant that he infused our research group not only with a steady stream of excellent scientific ideas but also with a certain intimacy. Stephen's Cambridge school was about black holes and the universe, yes, but I believe we learned even more from him about spirit. He taught us as much about courage and humility and how to live as he did about quantum cosmology.

Of course, our collaboration unfolded while Stephen was also busy being famous. But he left fame outside the walls of DAMTP. Reading the morning newspaper, I could come across a page-wide image of him driving through Ramallah, or floating in midair on a zero G flight, but once inside DAMTP he was just one of us, really, searching and struggling to understand the universe and its laws at the deepest level, and thoroughly enjoying this.

Stephen was a miracle. He embodied an extraordinary mix of seeking to understand some of the biggest questions in science with a lightheartedness, a certain irresistible playfulness that could pop up at every moment and wherever he was. One day he recklessly checked himself out of Papworth Hospital, to go and see a pantomime play. When it came to science, Stephen's lectures *had to* contain a few jokes.* Always. And for all his cryptic oracle-like talk, he too took pleasure in chitchatting. (Never mind that this took ages as well.)

Stephen's unique blend of wisdom and fun made certain there was nothing short of magic happening around him wherever he went. And, of course, it helped that he could never ever enter a room in a quiet, unobtrusive way.

WHEN I ARRIVED on his doorstep in June 1998, Stephen's quantum cosmology program was in full swing. The frenzy following the publication of *A Brief History of Time* had faded, the second string theory revolution was producing fantastic theoretical insights, and Stephen's team was buzzing with activity. Meanwhile, advances in telescope technology were transforming cosmology from a field rife with speculation into a quantitative science grounded in detailed observations spanning billions of years of cosmic evolution. This was cosmology's golden decade of discovery, when it felt as if the book of nature lay wide open for us to devour.

* To tell a joke in a smaller circle, with people looking over his shoulder and following every word on his screen, Stephen developed this really ingenious way of phrasing jokes so that it wouldn't be clear till the very last word whether he was conveying a profound insight or an ordinary joke.

The Cosmic Background Explorer satellite, or COBE for short, launched by NASA in 1989, had played a key role in our reading of the first few pages of cosmic history. One experiment on board COBE had established that the ancient cosmic microwave background radiation (CMB) had a near-perfect thermal spectrum, with a temperature of 2.725 Kelvin. But COBE had carried a second experiment, the differential microwave radiometer, designed to scan for minute differences in the temperature of the CMB radiation in different parts of the sky. This was an epic experiment. Cosmologists had known all along that the early universe couldn't have been exactly uniform, simply because the later universe isn't exactly uniform. Today we find matter aggregated into galaxies and clusters of galaxies. Had the universe started out as a perfectly uniform gas, this web of galaxies would never have formed and, since galaxies are the cosmic cradles of life, we wouldn't exist. By contrast, even the smallest density variations in the primeval plasma would over time have become amplified under the influence of gravity, potentially causing matter in the denser regions to clump and form cosmic structures. Calculations of the competing effects of expansion and gravitational clumping show that to grow galaxies in a period of about ten billion years, the baby universe must have had seed density contrasts of at least one part in a hundred thousand. Ever since the serendipitous discovery of the CMB in the mid-sixties, cosmologists had searched it for traces of these fluctuations. The COBE satellite was their last hope. Designed to reach this crucial level of sensitivity in the hunt for our cosmic roots, COBE was tapping into the basic consistency of the hot big bang theory.

To cosmologists' relief, COBE found precisely what it was looking for. Its data revealed that the early universe did, indeed, have slightly hotter and colder regions. While the temperature of the CMB is 2.7250 K *on average,* it tends to be 2.7249 K in one direction and 2.7251 K in another patch of the sky. "It's like seeing God," COBE's lead investigator euphorically declared at the press conference.

The faint microwave photons are some of the oldest we can hope to observe.[1] We cannot peer into earlier epochs with photon-gathering telescopes. Yet we are bound to wonder what gave rise to the tiny flickers in the primordial heat. The minuscule variations in the CMB radiation must, after all, be the outcomes of processes operating at earlier times still. Unfortunately, COBE had a spot of blurry vision and was unable to resolve the microwave background on scales less than about 10 degrees. This left cosmologists in the dark about the origin of the slightly hotter and colder blobs that it saw. But COBE did make cosmologists realize what a treasure of information is encoded in the ashes and smoke of the primeval fireball, if only they could read the small print in the background radiation. Ever since COBE, the faint microwave background has been the canvas on which modern cosmology projects its deepest questions.

And so it happened that at the close of the twentieth century, "golden" astronomical observations were finally beginning to decode the universe's birth certificate, realizing a vision Lemaître had formulated seventy years earlier:[2]

> *The evolution of the world can be compared to a display of fireworks*
>   *that has just ended;*
> *some few red wisps, ashes and smoke.*
> *Standing on a well-chilled cinder,*
> *we see the slow fadings of the suns,*
> *and we try to recall the vanished brilliance of the origin of worlds.*

EVER COMMITTED TO connecting cosmological theory with observation, Stephen, too, had high hopes that by painstakingly sifting through the ashes, cosmologists would be able to reconstruct the origins of the universe.

By the 1990s, Stephen had grown quite attached to his no-boundary hypothesis. The way it ingeniously circumvented the age-old paradoxes associated with a beginning of everything had an irresistible charm. To Hawking, it also had the ring of truth. Ample evidence suggests he regarded it as his greatest discovery indeed.[3] But no matter how elegant or beautiful a cosmological theory, its real test lies in its predictions, and Hawking was the first to emphasize this. Suppose the universe was indeed born "from nothing," out of a spherical nugget of pure space. What exactly would the speckled CMB map look like? An intriguing question, and one that now topped Stephen's agenda. But to answer this we must first return to cosmic inflation, the idea that the universe underwent a brief burst of superfast expansion early on.

The theory of cosmic inflation was pioneered in the early 1980s by the theoretical physicists Alan Guth, Andrei Linde, Paul Steinhardt, and Andreas Albrecht. It counts as the most important refinement of the hot big bang model since its inception. Originally, inflation was thought of as a brief transient phase very early on in the universe's history when gravity would have been strongly repulsive, driving an intense burst of extreme expansion. Inflation's pioneers envisaged that, in less than a split second, the observable universe would have swollen by a stupendous factor of 10^{30}. That corresponds roughly to the difference in scale between an atom and the Milky Way.

This expansion boom appealed to theorists because it could neatly explain the puzzle that I discussed in chapter 3: Why is the universe so smooth and uniform on the largest scales? A short shot of superfast expansion would mean that even the most distant regions in today's observable universe would have been close together initially, at the onset of inflation, lying nicely within each other's horizon. Looking at figure 19, what happens is that even the briefest burst of superfast inflation would push the big bang singularity much further down, thereby creating a single interconnected environment covering our entire past light cone. The entire observable universe would have thus had a common causal origin, from which it could have emerged nearly the same everywhere.

Yet on the face of it, the staggering numbers behind inflation sound crazy. To put these in perspective, the immense ballooning of space during that brief instant of inflation would far exceed the universe's total expansion factor during the subsequent 13.8 billion years! What strange form of matter could possibly cause space to stretch in such a dramatic manner? Inflation's pioneers proposed that scalar fields could have been responsible. Such fields are exotic forms of matter that can behave as invisible, space-filling substances, similar to electric and magnetic fields, but that are even simpler in that they only have a value, not a direction, at each point in space. One well-known scalar field is the Higgs field, the capstone of the Standard Model of particle physics that was discovered at CERN* in 2012. Theoretical extensions of the Standard Model typically contain a large number of scalar fields, some of which may be part of the dark matter in the universe. The one responsible for inflation is aptly—

---

* European Organization for Nuclear Research

though perhaps confusingly—called the inflaton field. The inflaton field is a hypothetical field, hitherto not found at CERN or anywhere else on Earth, which, inflationary theory predicts, would have briefly boosted the early universe's expansion to truly mind-boggling scales.

But why are scalar fields such powerful sources of repulsive antigravity? Scalar fields feature on the right-hand side of the Einstein equation (see page [42]), together with all other forms of matter. Unlike normal matter, however, scalar fields share some important properties with the cosmological constant, Einstein's λ term. Like the cosmological constant, scalar fields spread out uniformly fill space not only with positive energy, generating attractive gravity, but also with negative pressure, or tension, which causes antigravity. And the antigravity of scalar fields, it turns out, beats their gravity, which is why they accelerate the expansion, unlike all other forms of matter. Furthermore, inflation feeds on the expanding space. Whereas familiar matter loses energy to an expanding space, the negative pressure with which the inflaton field infuses the universe means that, just like the cosmological constant, it doesn't quite dilute but actually gains energy from the expansion.[4]

When, back in 1917, Einstein added the cosmological constant to his theory, he finely tuned its value so that its repulsion perfectly balanced the gravitational attraction of matter, and the universe stood still. Sixty years

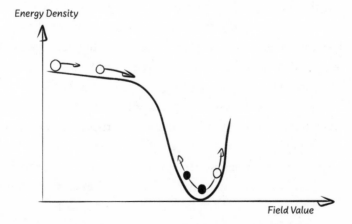

FIGURE 27. *The energy density contained in the inflaton field on the vertical axis, for different values of the field on the horizontal axis. While the universe inflates, the field tends to roll down toward a valley in its energy landscape.*

on, inflation's pioneers went much further and envisioned that the inflaton field's antigravity, during a brief instant in the very early universe, would have far overpowered all sources of attractive gravity and provided the big bang with a real bang—a brief burst of enormous cosmic expansion.

Figure 27 shows inflationary theory at work. The curve represents the energy density contained in a (hypothetical) inflaton field for different values of the field. The height of this curve indicates the strength of the inflaton's antigravity. Inflationary cosmology imagines that in the earliest stages of the universe, there was a small region of space in which the inflaton field somehow wound up high on the plateau of its energy curve. This would have caused that patch of space to inflate, while the inflaton field inside it would have gently rolled down toward the valley in its energy landscape. Once the inflaton tumbled down to its minimum energy state, inflation would have run out of fuel. The cosmic growth spurt would have drawn to a close and the universe would have transitioned to a much more modest pace of expansion. Hence even though they both lead to repulsive gravity, an inflaton field differs from a cosmological constant in one important respect: A cosmological constant is, obviously, constant whereas the value of an inflaton field can change over time. This inflaton quality makes it possible to turn on and switch off bursts of rapid expansion—a key property exploited by inflationary theorists.

Now, at the end of the inflationary burst, the huge amount of energy stored in the inflaton field would have had to go somewhere, and that somewhere is heat. When inflation ceased, the tumbling inflaton would have filled the universe with hot radiation. Some of this heat energy would subsequently have turned into matter, because Einstein's formula $E = mc^2$ tells us that as long as there is enough energy (E) to pay for the mass (m) of a particle, the way lies open for highly energetic particles of radiation (photons) to be transformed into massive matter particles. At the end of the flash of inflation, the intense energy released may well have heated the universe to around a thousand trillion trillion degrees, more than sufficient to create the 10^{50} tons of matter contained in the observable universe.

SO, INFLATION PRODUCES a fantastically large and uniform universe in less than the blink of an eye. But what about the vital flickers in the CMB that COBE spotted? Does inflation produce a universe that is almost, but not quite perfectly, smooth?

As a matter of fact, it does. Like all physical fields, the inflaton is a quantum field. Heisenberg's uncertainty principle says that it too must be subject to irreducible quantum fuzziness. Much like particles, this means that the more precisely we pin down the value of a field at a particular location, the less precisely its rate of change at that location can be known. But if a field's rate of change is somewhat uncertain, we cannot know what its exact value will be a moment later. Quantum fields, that is, assume a strange jittering mixture of many different rates of change and values, much like the many paths making up the wave function of a particle.

Normally such quantum jitters are extraordinarily small and confined to microscopic scales. But a surge of cosmic inflation is anything but normal. To their amazement, theorists studying inflation soon realized that the enormous burst of expansion they were envisioning would amplify microscopic quantum fluctuations and stretch these to macroscopic wavelike variations. Even if the inflaton started out with the bare minimum level of jitters allowed for by the uncertainty principle, a burst of inflationary expansion would transform these into macroscopic wobbles, superimposing on the overall smoothness of the expanding universe a wavy pattern of variations in the field, much like ripples on the surface of an otherwise smooth lake.

Crucially, when inflation ends and the inflaton liberates its energy into a burst of heat, the hot gas that fills the newborn universe inherits these variations. So the upshot is that any universe that emerges from inflation comes equipped with small irregularities in the temperature of the radiation and in the density of the matter. Under the slowing cosmological expansion that ensues, more and more of these primeval ripples would roll within our cosmological horizon and become visible, somewhat like waves arriving on the shore. The fluctuations in the radiation temperature would be there for us to admire. They would show up as hot and cold spots in the CMB as we compare its temperature in different directions in the sky. But the density variations in the matter would become important, for they can seed the growth of galaxies. Regions where there is less density initially would expand faster and empty out. Regions with more matter would start drawing in even more material from their surroundings, enhancing the density contrast and, in this way, sculpt the large-scale web of galaxies we see today.

In the summer of 1982 Stephen and Gary Gibbons brought the principal inflationary theorists together in Cambridge for what, years later, he fondly remembered as a *real* workshop. The Very Early Universe was funded by the Nuffield Foundation, a charitable organization set up in the 1940s by automobile magnate William Morris, Lord Nuffield.* For days on end Stephen and his colleagues debated what the key characteristic properties were of the primordial variations generated through inflation. By the end of the workshop they agreed that, lo and behold, a burst of inflation would imprint a difficult to discern yet distinctive, clearly recognizable signature in the flickering CMB fluctuations.[5] That is, the theorists at the Nuffield workshop had identified a smoking gun from inflation that should be out there for us to discover if only we meticulously scanned the microwave sky. Their findings rank as one of the most spectacular predictions of theoretical cosmology and perhaps all of science. The wavelike relics of inflation, frozen in and preserved in the CMB with a stunning mathematical precision, are undoubtedly among the oldest fossils one could ever hope to identify.

For good reasons, the Nuffield gathering became legendary. The 1982 Nuffield workshop meant to cosmology what the 1911 Solvay Council had meant to atomic physics. Its results amounted to a coming-of-age for the study of the very early universe. The predictions of inflationary theory clearly demonstrated that quantum mechanics had far-reaching implications not only for the microscopic world but also for our observations of the universe on the very largest scales. Much as the 1911 Solvay Council had marked the moment quantum mechanics was understood to be central to the atomic world, the 1982 Nuffield workshop showed that quantum mechanics was essential to cosmology. Inflationary theory said that the hot and cold spots in the CMB are primeval quantum fuzz, magnified and writ large across the cosmic sky. And even more, it predicted that an advanced version of COBE capable of taking a sharp image of the CMB

---

* In fact, this was the second Nuffield meeting Stephen hosted in Cambridge. The first one, on supergravity, "held to be an instructive way to spend four weeks," as Stephen jokingly summarized it, was equally memorable, with the blackboard that creatively illustrated the proceedings decorating Stephen's office till the end of his life (see insert, plate 10).

speckles should be able to verify all this. Such a picture would establish a magnificent arc linking our cosmological observations today to microscopic quantum jitters no more than 10^{-32} seconds after the big bang.

Stephen didn't hide his excitement about the results of the workshop, writing, "The inflationary hypothesis has the great advantage that it makes predictions about the present density of the universe and about the spectrum of departures from spatial uniformity. It should be possible to test these in the fairly near future and either falsify the hypothesis, or strengthen it."[6]

Blobs and speckles in the CMB generated by inflation are the cosmological analogue of Hawking radiation from black holes, another fascinating connection between black holes and the big bang. I mentioned earlier that Hawking radiation originates in quantum jitters of matter fields in the vicinity of black holes. These jitters give rise to pairs of particles that pop into existence, persist for a short while, and disappear again, like a pair of dolphins rising briefly above the surface of the ocean before diving down again. Physicists call these *virtual particles* because, unlike real particles, they don't live long enough to be detected by a particle detector. Near the horizon of a black hole, however, virtual particles can become real. This is because one member of a virtual pair can fall into the black hole, leaving the other particle free to escape into the distant universe where it shows up as faint radiation given out by the black hole.[7] The story of an inflating universe is like that of a black hole turned inside out: Rapid inflationary expansion amplifies quantum jitters associated with the cosmological horizon that surrounds us, which makes the universe twinkle ever so slightly in the microwave frequency band. Inflation predicts that we are immersed in a cosmic bath of Hawking radiation.

DESPITE THE ODDS, Stephen lived to see the detailed observations of the CMB radiation carried out and saw to his great satisfaction that the observed pattern of variations matched the rhythm of inflation indeed.

In the summer of 2009 the European Space Agency launched the Planck satellite, which succeeded in gathering ancient microwave photons for nearly fifteen months. Satellites do this better than Earth-based telescopes because they don't have to peer through our atmosphere and can scan the full sky. The Planck satellite recorded the temperature and polarization of CMB photons reaching us from millions of different directions in space. The Planck army of astronomers then compiled an exquisitely detailed map of the microwave sky, transforming COBE's hazy view into an unprecedentedly sharp image. Plate 9 (see insert) shows this image and encourages you to imagine the CMB sky as the giant, most distant sphere that it is, a cosmic horizon all around us with Earth at the center of the ball. The CMB map that I showed earlier in figure 2 is a projection of this sphere onto a flat sheet, just as how one makes a map of the world.

At first glance the spots and speckles on the CMB sphere appear random, but a closer look has revealed that coded into these few millions of pixels are the long-sought signature variations of a primordial surge of inflation.

Figure 28 represents this inflationary pattern of fluctuations, showing the expected strength of temperature differences against the angular scale of the sky over which they are measured. We see that the level of variations oscillates and decays away toward small angles, like the ringing of a bell. The agreement between the Planck satellite's observational data—the dots—and

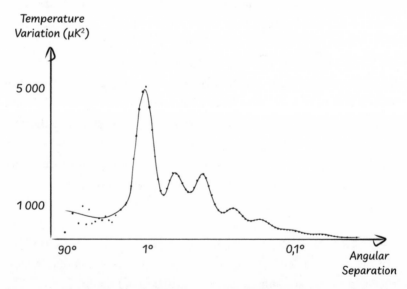

FIGURE 28. *The expected level of CMB temperature differences on the vertical axis is shown against the angular separation between two points in the sky on the horizontal axis. Larger angles are to the left, smaller angles to the right. The solid line indicates the prediction of inflationary theory. The dots are the Planck satellite's data points. The data line up near-perfectly with the oscillating pattern predicted by the theory.*

the theoretical predictions—the curve—is stunning. This wavy pattern of variations has become one of the iconic images of modern cosmology. It is widely recognized as the first strong evidence that our deepest origins lie in quantum jitters amplified and stretched in a brief surge of primeval inflation. Planck (the satellite) truly lived up to its eponymous genius.

What's more, the oscillations in the level of CMB variations also tell us something about the makeup of the universe today, and even about its future. This is because the finer details of the spectrum of fluctuations depend not only on their inflationary origin but also on the universe's geometry throughout its evolution. Using Einstein's theory to relate the geometric shape of spacetime to its content, Planck's precision data have allowed physicists to learn a great deal about the universe's composition.

Take the location of the first peak in figure 28, which occurs at a separation angle in the sky of about 1 degree (for comparison, the full moon extends roughly half a degree). The location of this peak shows that the spatial shape of the observable universe is hardly curved. So, if the three dimensions of space form a hypersphere, then this must be a phenomenally large one that looks flat even on the scale of our cosmological horizon, much as Earth looks flat around us.

The height of the second peak demonstrates that ordinary visible matter, like protons and neutrons, accounts for only about 5 percent of the total content of the universe today. The third peak comes to the rescue and shows that the universe also contains about 25 percent dark matter, mysterious types of particles that barely interact with ordinary matter or light, if they interact at all.[8] Dark matter has nevertheless played a crucial role in the history of the universe, providing the additional gravitational pull needed for the tiny seeds in the primordial gas to grow into a web of galaxies. You can think of dark matter as the cosmic backbone that has guided the organization of visible matter into the large-scale structures that make our universe habitable at all.

So, from the heights and locations of the first three peaks in Planck's oscillating curve we arrive at the somewhat unsettling conclusion that about 70 percent of the content of today's universe isn't matter at all (see figure 29). On the contrary, the largest piece of the cosmic cake is invisible, antigravitating dark energy, responsible for the surge in expansion in the universe's recent era. This reading of the CMB corroborates the spec-

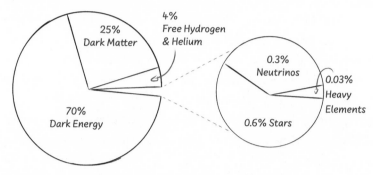

FIGURE 29. *A pie chart showing the matter and energy budget of the universe today. The bulk of it consists of dark energy that has been driving the acceleration of the expansion of the universe in the last few billion years. The remainder is mostly in the form of nonatomic, dark matter composed of unknown particles. Only a small fraction, roughly 5 percent, consists of ordinary, familiar matter and radiation.*

tacular discovery by two teams of astronomers who, in 1998, from observations of light emitted by distant exploding stars, found that the expansion of space had been speeding up in the last few billion years.[9]

Now, if the dark energy is really nothing more than Einstein's λ term, an energy associated with empty space, then it will have a dramatic effect on the universe's far future. Once a cosmological constant takes control, it is there to stay, for unlike an inflaton field, you can't switch off a constant. So if there is a truly constant cosmological constant out there, then the acceleration of space may well continue forever. In this future universe the formation of new stars and galaxies would come to a halt, the existing galaxies would either collide or gradually disappear beyond one another's horizon, and the night sky would slowly turn dark,[10] depriving future astronomers of much pleasure.

Today, a wide range of astronomical observations have fallen in line and a great number of checks and cross-checks of the cosmological model summarized in figures 28 and 29 have been performed. Physicists are now confident they know the composition of the observable universe and its expansion history to a high level of precision. The concordant picture that has emerged in the wake of the golden decade of cosmology is strikingly similar to the one Lemaître sketched almost ninety years ago, with a brief inflationary burst, followed by an extended near-pause in expansion, and, finally, a transition to a much milder phase of acceleration (see insert, plate 3).

According to James Peebles, who received the 2019 Nobel Prize in Physics for his crucial role in putting together a coherent model that accounts for our cosmological history, "So far there are no clouds on the horizon."[11]

THERE IS ONE key prediction of inflation, however, that has remained elusive. These are the primordial gravitational waves. Because an inflationary surge of expansion amplifies all quantum jitters, including those of space itself, it also generates a certain level of gravitational waves. These ripples of space are known as primordial gravitational waves, to distinguish them from gravitational waves produced much later in the collisions of black holes, neutron stars, or galaxies.

Primordial gravitational waves from inflation would have expanded in sync with the universe ever since its birth. By now they would have exceedingly long wavelengths that don't fit in the iconic L-shaped observatories on the surface of Earth. But the mere existence of gravitational ripples from inflation, undulating throughout space, would affect the polarization of the microwave background photons, since these would have been traveling through a slightly rippling geometry for 13.8 billion years before hitting the disks of our telescopes. And even though the expected level of primordial gravitational waves is relatively low, even by gravitational-wave standards, inflationary theorists believe their polarizing effect on the CMB radiation should be detectable.

Unfortunately Planck wasn't equipped with great polarimeters. A later experiment that was designed to measure the polarization of the CMB, operating from Antarctica at the Amundsen-Scott South Pole Station, did find the kind of polarization expected from inflation. However, a close scrutiny of its data revealed that this polarization could be traced to the interference of galactic dust. But cosmologists aren't giving up. New satellite missions are being envisaged to search for the imprint of primordial gravitational waves on the cosmic microwave background sky. Even though gravitational waves from inflation may not contain a wealth of information, their observation, even indirectly, would be an exhilarating discovery. Not only would it consolidate the theory of inflation, it would also be the first tangible evidence that the field of spacetime really has quantum roots just like all known matter fields.

Stephen, too, had pinned his hopes on the detection of gravitational waves from inflation. At the time of his death, he was working on a paper in which he hoped to refine the predictions of inflationary theory for the exact level of primordial gravitational waves. He had a great deal at stake on this project, for the inflationary stretching of quantum jitters to the macroscopic is the cosmological counterpart of Hawking radiation from black holes. Most physicists would agree that, indeed, if the footprint of primordial gravitational waves were found, it would constitute convincing evidence—albeit indirectly—for Hawking radiation.

THE THEORY OF inflation offers an extraordinarily successful description of a brief but crucial instant in the universe's genesis. Even though the exact nature of the inflaton remains obscure and the primordial gravitational ripples elusive, the detailed observations of its distinctive ringing pattern of temperature variations means that most cosmologists are sold on inflation. Inflation feels right and looks right. But this also means that the question of how it could have begun becomes of paramount importance. For when we attempt to explain the very early universe, we must be careful not to trade one puzzle for another. No matter how theoretically appealing inflation may be, if a significant surge were impossible to ignite in the first place, we would be left empty-handed and inflation would be rendered irrelevant as a physical model of the early universe. That's how science works.

So what does it take to initiate inflation? How could the inflaton field have wound up high on its energy hill in the first place? This is where the no-boundary proposal comes in. Remarkably, the no-boundary hypothesis *predicts* that the universe originated in a burst of inflation. Mathematically this is because the rounded shape at the bottom of spacetime in the no-boundary process of creation requires the same sort of exotic scalar matter, exerting negative pressure, as inflation does. In the context of real-time classical cosmology, matter with negative pressure can cause rapid runaway expansion—inflation. In the context of imaginary-time quantum cosmology, the very same negative pressure makes it possible to close the bottom of spacetime smoothly like a sphere. So no-boundary creation and inflationary expansion are twin processes that go hand in hand. They reinforce

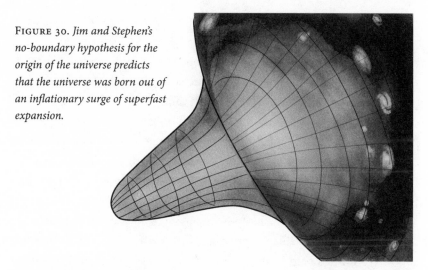

FIGURE 30. *Jim and Stephen's no-boundary hypothesis for the origin of the universe predicts that the universe was born out of an inflationary surge of superfast expansion.*

each other, the former being the quantum completion of the latter (see figure 30). Physically this means that if the universe were created from nothing—according to the rules of the no-boundary theory—then the chance of it following most of the possible expansion histories would be utterly negligible. But there would be one particular family of trajectories that is much more probable than others. These are the expansion paths where the universe comes into existence with a brief burst of inflationary expansion, before subsequently slowing down.

I mentioned earlier in chapter 1 that at any level of evolution, determinism shapes only the most general structural trends. Only the coarsest-grained properties can typically be predicted in advance. According to the no-boundary hypothesis, some form of inflation would be such a structural property of cosmological evolution.

The discovery of this intriguing resonance between the no-boundary hypothesis and inflation was a thrilling experience for more than one generation of Hawking's students. And it has far-reaching implications. Its pioneers conceived of cosmic inflation as an intermediate, transient phase in a preexisting universe. Its quantum completion, however, suggests that inflation *is* the beginning. In the no-boundary proposal, inflation becomes an integral part of the quantum process that gives rise to a classical fabric of spacetime in the first place. So the no-boundary proposal takes inflation to a higher level, and ties it to the very existence of spacetime. The origin of inflation wouldn't be a mysterious fluke, or the result of the

"Finger of God" placing the inflaton up the hill, but a cosmic imperative in order for the universe to exist at all.

Yet, there is a problem: The no-boundary proposal predicts the most minimal inflationary burst possible. The strength of the initial surge of expansion is determined by the starting value of the inflaton field. Universes in which the inflaton starts out high on its energy hill in figure 27 undergo an enormous burst of inflation. They end up bigger and with enough matter to form billions of galaxies. These are much like the universe we observe. By contrast, universes in which the inflaton starts out near the lower edge of its energy plateau emerge with just a whisper of inflation. Such universes end up being nearly empty, devoid of galaxies, and they may even recollapse into a big crunch. They are not at all like our universe. Unfortunately, the no-boundary theory, taken at face value, picks out precisely the latter universes. The theory appears to say that we should find ourselves in a universe where we can't be. Not surprisingly, then, most physicists have found it hard to seriously engage with no-boundary creation. And this has been the elephant in the room ever since Jim and Stephen put forward their model of cosmogenesis.

LET'S TAKE A closer look at this elephant. The mystery of how inflation started is tied closely to that of the arrow of time—another obvious property of the world. It is abundantly clear from everyday experience that there is a definite direction to the way things happen. Eggs break but don't unbreak. People grow older, not younger. Stars collapse into black holes but don't come out again. Above all, we remember the past but not the future. This directionality, this arrow of time, is one of the most powerful and universal organizational principles behind the workings of the physical world. We simply never come across unsplattering eggs or black holes ejecting stars. But how did time acquire such a robust arrow?

In ancient times, people held a teleological view of time's arrow. The obvious directionality of the way many things happen matched seamlessly Aristotle's idea that the workings of nature were directed by a "Final Cause."

Today, by contrast, we understand that the arrow of time effectively arises from the tendency of disorder to increase. Think of your office, or your bedroom, which tends to get messier unless you put in a real effort to maintain order. This is because there are many more ways for an office to be messy than there are for it to be tidy. Or take the pieces of a jigsaw. If you were to shake a jigsaw puzzle in its box, you'd be very surprised to find the pieces perfectly arranged to reproduce the picture on the cover. Again this is because there are so many more configurations corresponding to a disordered jigsaw than to an ordered one. These examples illustrate a universal property of physical systems: There are far more ways to be messy than to be orderly, and this is why physical systems made up of many components tend to evolve toward greater disorder.

Scientists measure the amount of disorder in a physical system by its *entropy*, a concept that goes back to the nineteenth-century Austrian physicist Ludwig Boltzmann. High entropy means that a system is in a highly disordered state, and low entropy is one that is highly ordered. The tendency of complex physical systems to evolve toward states of higher entropy implies a quasi-universal arrow that is known as the second law of thermodynamics. This entropic arrow is the source that underpins the arrow of time we experience.

But here's the mystery. Obviously entropy can increase only if it starts out low. So why was the entropy lower yesterday than today? How come we have unbroken low-entropy eggs available to make an omelet? Eggs come from chickens, which are low-entropy systems on farms that are themselves part of a low-entropy biosphere. To sustain itself the Earth's biosphere draws on the energy of the sun. So where did the low-entropy sun come from? The sun arose out of a very low-entropy cloud of gas that nearly five billion years ago collapsed and that was itself the remnant of previous generations of stars. So what about the extremely low-entropy cloud of gas that is responsible for the first generation of stars? That cloud can ultimately be traced to the tiny variations in the density of the hot gas filling the early universe, whose seeds might have been planted during inflation.

The universe at the end of inflation must have had an extraordinarily low entropy indeed.

So this chicken-and-egg story is telling us something profound. It tells us that the ultimate source of order, the reason we have unbroken low-entropy eggs today, has to do with our big bang origins. Nearly fourteen billion years ago the universe started off in an incredibly ordered manner, and we have been riding its natural evolution toward greater disorder ever since. The arrow of time that distinguishes the past from the future, arguably the most basic element of our experience, finds its origin in the extremely ordered, low-entropy state of the primordial universe. This is, perhaps, the most enigmatic biofriendly property of all. How did the universe come into being in such a pristine low-entropy state? Did the burst of inflation somehow cleverly lower the entropy of the very early universe, violating the second law of thermodynamics? It didn't. The entropy increased during inflation (albeit more slowly than it might have) and continued to do so as the cosmos evolved.

This point has been made most forcefully by Penrose, who for this reason calls inflation a "fantasy." In order for inflation to start, the inflaton field must have been in an exceedingly low-entropy state, perched high on its energy curve, which Penrose regards as an unreasonably fine-tuned initial condition. But the embedding of inflationary theory in quantum cosmology has the potential to address Penrose's concern. As a theory that unifies dynamics with initial conditions, the no-boundary hypothesis has a certain time asymmetry built in, with a smooth inflationary birth at one end of cosmological history and an open-ended, disordered state at the other end. The arrow of time implied by the no-boundary proposal, however, appears nowhere near strong enough to breathe life into the universe. The theory places the inflaton just slightly up the hill, in a state of intermediate entropy. Jim and Stephen's scheme would seem to create the universe not with a bang but with a whisper. The no-boundary hypothesis may be elegant, profound, and beautiful, but it doesn't work. The second law of thermodynamics wins.

THERE WAS JUST one glimmer of hope. This hope was to be found deep in the quantum roots of the no-boundary hypothesis. You see, as a theory of the wave function of the universe, the no-boundary proposal doesn't uniquely select the absolute minimum amount of inflation but describes a somewhat fuzzy origin of the universe. Just as the wave function of a single electron encompasses an amalgam of electron trajectories, each with a certain amplitude, the no-boundary wave function slightly spreads over an array of inflationary universes, each with a different inflaton starting value. A quantum universe, that is, isn't just a single expanding space, but different possible expansion histories living in superposition, very much in the way Doc Brown explained it to Marty on the blackboard in *Back to the Future*.

To get a feel for this abstract quantum cosmos, consider again the expanding circular universe. Figure 30 depicted the no-boundary creation of a one-dimensional circle-shaped universe. But this was only one specific history of expansion, representing just a sliver of the no-boundary wave. The circle in figure 30 is surfing one particular wave crest in a much larger quantum reality. To picture the no-boundary wave in its entirety, one would have to imagine a collection of circles, each one expanding in its own characteristic way. I attempt to evoke this mind-boggling quantum cosmos in figure 31. The collection of expansion histories shown here in some sense coexist in the no-boundary wave, a striking manifestation of the fuzzy nature of spacetime in a quantum world.

This coexistence of universes is both intriguing and confusing. In classical relativity theory, one spacetime has nothing to do with another. For example, every curve in Lemaître's iconic diagram in plate 1 of the insert describes a separate universe, and there is nothing whatsoever in Einstein's theory that gives weight to one over another. Not so in quantum cosmology, where Stephen's wave function operates in the vast arena of all possible cosmic histories. Just as the quantum mechanics of an electron particle unites different possible trajectories in one entity—the electron wave function—the no-boundary wave function brings together different possible expanding universes under a single umbrella. Precisely herein lies its capacity to yield further theoretical insights into the question "Which curve should be ours?," so central to the riddle of design.

Interestingly, this bundling together also means that the wave function as a whole doesn't change in time. Indeed, I did not indicate an overarching notion of time in figure 31, a universal clock relative to which the entire collection of expanding universes would evolve. Time in quantum cosmology loses its meaning as a fundamental organizing principle.[12] Instead, a sensible notion of time emerges only as an intrinsic quality within each individual expanding space. The reason is that a measure of time always involves a change of one physical property relative to another. As a clock within our own universe, for example, we could use the monotonic cooling of the cosmic background radiation with expansion (although this wouldn't be a practical unit of time to schedule your meetings). But

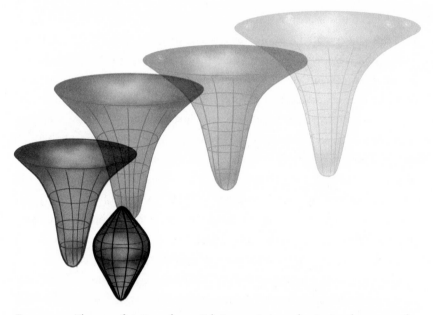

FIGURE 31. *The wave function of a particle in quantum mechanics involves an amalgam of all possible particle paths (see figure 21). Likewise, the wave function of the universe in quantum cosmology describes a collection of all possible expansion histories. The shape of Hawking's no-boundary wave function, taken at face value, is dominated by universes that undergo a minor burst of inflation and rapidly collapse again. Universes with a strong burst of inflation that form galaxies and become habitable aren't completely ruled out by the theory, but they reside far in the tail of the wave function. They are barely visible in the theory.*

the evolution of the temperature of the CMB in one spacetime is obviously of no use as a clock in a different spacetime.

Sadly, however, the intrinsic spread of the no-boundary wave isn't nearly broad enough to cover any of the habitable universes with a strong surge of inflation. As a probability wave over the strength of the inflationary burst, the no-boundary wave has an extraordinarily sharp peak at the minimal-inflation universe, and only an exponentially small tail extending toward universes with a more significant burst of inflation. So while the no-boundary proposal profoundly resonates with inflation, relying on the same sort of negative pressure to create spacetime, it also implies that the most minimal amount of inflation, barely enough for the universe to exist, is overwhelmingly more likely than the more interesting expansion histories with more inflation.

This state of affairs is puzzling. Should we expect to live in the most probable universe? More important still, should we disregard a theory of the universal wave function because the universe we observe lies in the far, far tail of the probability wave? Remember, it takes matter in the form of atoms for there to be observers wondering which universe they're in. If the most probable universe in a cosmological theory is empty and lifeless, then we shouldn't be surprised that this is not where we find ourselves. Moreover, if certain properties of the universe are indispensable for there to be life, such as galaxies, we should not simply dismiss a universal wave function that predicts that the most probable universe has no galaxies. What matters is not what is most probable in the theory but what is most probable to be observed. Cosmological histories that don't produce observers don't quite count when we compare our theories to our observations.

Reasoning along these lines, in 1997, Stephen and Neil Turok attempted to rescue the no-boundary theory by augmenting it with the anthropic condition that "we" should exist in the universe.[13] However, they found that that stipulation barely made a difference; the theory complemented with the anthropic principle ended up predicting a universe with just one galaxy—ours—and nothing in any way resembling a universe teeming with galaxies like the one we observe. This disappointing result appears to have made a strong impression on Turok at the time, who radically switched gears and went on to devise new ways to avoid a beginning altogether. Stephen, however, held on to the no-boundary proposal; in hindsight, he was just getting started.

MEANWHILE, A RIVAL vision of the origin of inflation had emerged from the work of Andrei Linde and Alexander Vilenkin, a Ukrainian-born American cosmologist and a deep thinker of few words, working at Tufts University. Their proposal was so radical, its implications so mind-blowing, that it has riveted the cosmology community ever since: the multiverse.

Linde and Vilenkin turned the problem of igniting inflation upside down. They argued that inflation has forever been the universe's default state and that it is, in fact, hard to stop inflation. Inflationary expansion, they suggested, is eternal by nature.[14] Their reasoning involved the same kind of quantum jitters that grow into the seeds of galaxies during inflation but now conceived on much larger scales, scales far larger than our cosmological horizon. If inflation created such extraordinarily long--wavelength ripples, then the strength of the inflaton field would fluctuate across these vast distances. In some regions the fluctuations would help the inflaton roll down and bring inflation to a close, giving rise to a hot big bang followed by slow expansion. However, in faraway patches where the inflaton experienced a jittering jump that strengthened it, inflation would actually pick up. Now, Linde and Vilenkin argued that even though such regions may be rare, their higher rate of inflation means they generate so much volume that there will always be some regions where the strengthening jumps win and the inflaton remains hovering high on its energy plateau. From a global perspective, then, inflation would be much like a raging pandemic, a self-sustainable process in which inflating regions spawn further inflating regions that in turn produce local big bangs or even more inflation, and so on *ad infinitum*.

Obviously, the idea that inflation is eternal yields a radically different view of our distant past. The origin of inflation would be that there is no origin. Instead of a brief surge of primeval expansion tied to how space-time itself comes into being, inflation would be a perpetual and inexhaustible universe-generating mechanism. "The universe as a whole is a self-reproducing system," Linde wrote, "that exists with no end, and possibly no beginning."[15] The entire observable universe would be only an island universe in a much bigger space. Globally the cosmos would be a complicated superstructure—a multiverse. Within a single island region, quantum jitters stretched to cosmic scales would seed the growth of galaxies. But those jitters stretched to much larger scales would generate other shores altogether. If we could somehow look at the cosmos from the outside, we would see an elaborate cosmic quilt of slowly expanding islands, patches where the end of inflation spurred a cycle of evolution, embedded in a giant, possibly infinite inflating space. Some islands would contain a web of galaxies stretching as far as the James Webb telescope's eye could see. Others, where inflation happened to end abruptly, would barely have any matter available to form galactic structures. It would be completely impossible to travel from one island universe to another, even in principle, because the rapid expansion of the inflating ocean would make it physically impossible, even for light, to cross the widening gulf separating different islands. For all practical purposes, then, each island would behave like a separate universe.

This is a truly bewildering picture of physical reality. It reminds one of Thomas Wright's never-ending universe. Wright, an eighteenth-century clockmaker, architect, and self-taught astronomer from Durham in the north of England, was ahead of his time when he imagined the Milky Way galaxy as one of an endless number of galaxies, each one containing a great collection of stars. His drawings of a seemingly infinite space bubbling with spherical galaxies bear a striking resemblance to some of the visualizations of an inflationary multiverse (see insert, plate 7). Wright's never-ending universe fascinated Immanuel Kant, who spoke of galaxies as "island universes." Wright's and Kant's speculations were an important step in coming to terms with the idea of a bigger universe but did not win acceptance until 1925, when Hubble discovered that the spiral nebulae in the sky are separate galaxies indeed. However, Hubble's enlargement pales in comparison with that implied by the infinite multiverse envisioned by these inflationary theorists.

And the complicated fractal-like cosmography of the multiverse spooks us. In an eternally inflating multiverse, you will eventually find an island universe that has a galaxy that looks like an exact copy of the Milky Way, with a solar system that's just like ours and an identical house in an identical street where your doppelgänger is reading these words. Moreover, there'd be not just one such copy but infinitely many. I tried this idea out on our youngest daughter Salomé the other day. She firmly objected.

DURING A DELIGHTFUL dinner in Cambridge celebrating Stephen's sixtieth birthday—Stephen had a real knack for throwing parties—Andrei Linde recalled his first meeting with Stephen in a way only a Russian physicist could. It was in 1981 in Moscow, where Stephen was scheduled to give a lecture on inflation to an audience of distinguished Russian physicists at the Sternberg Astronomical Institute. At the time Stephen could still speak, but since his voice was difficult to understand he would often lecture by having one of his students repeat his words. For his lecture at the Sternberg Institute, a two-step process was set up in which a young student fluent in English and Russian, Linde, was asked to interpret into Russian what Stephen's student said Stephen had said. Having co-discovered inflation, Linde knew the subject well, and, being Russian, he could not resist elaborating, extensively, on Stephen's words. For a while all went well: Stephen said something, Stephen's student repeated, and Linde would expound. Then Stephen began to criticize Linde's model of inflation. For the remainder of Stephen's talk, Linde found himself in the unfortunate position of having to explain to the Russian physics elite why the world's foremost cosmologist thought his theory of inflation was plain wrong. It was the beginning of a lifelong friendship, Linde recalled. But it was also the birth of *the* major controversy that has gripped the field of theoretical cosmology ever since.

Linde versus Hawking on the origin of inflation was in some ways a replay of Hoyle versus Lemaître, this time in semiclassical cosmology, the amalgam of classical and quantum physics that both Linde and Hawking were employing. In the 1950s Hoyle had tried to hold on to the idea of a steady-state universe by invoking the continual creation of matter to fill the voids left behind by galaxies moving apart. Lemaître, by contrast, had fully embraced the idea of an evolving universe that was dramatically different in the far past. We go from Hoyle to Linde, from classical to semiclassical cosmology, by replacing galaxies with universes. For in a similar vein, the continual creation of island universes in the multiverse spawned by eternal inflation gives rise to some sort of steady state, globally, on the much larger scale of the multiverse. The enigma of the ultimate cause of inflation—and indeed whether it ever had a beginning—would seem to evaporate in an eternally inflating multiverse.[16] By contrast, there is not a trace of a steady state in Hawking's no-boundary cosmology. Quite the contrary, Hawking takes Lemaître's idea of cosmic evolution all the way to the extreme by bending time into space at the "onset" of inflation. Whereas multiverse cosmology presumes the stable backdrop of an eternally inflating space in which everything happens, the no-boundary proposal holds that quantum mechanics becomes so fundamentally important in the very early universe that it washes out even that backdrop—the very fabric of spacetime.

Stephen felt that the idea of an eternally inflating multiverse was an excessive extension of physical reality that was neither justified nor relevant for anything we can ever hope to observe. Andrei rallied against the no-boundary hypothesis on the grounds that it predicted no observers at all. A no-boundary origin selected the feeblest possible wisp of inflation, giving birth to an empty, lifeless cosmos. Linde's eternal inflation amounted to the strongest burst of inflation one can imagine, spawning not one but infinitely many universes and observers. While the no-boundary proposal said we shouldn't exist, eternal inflation saddled us with an identity crisis. And so it happened that cosmology emerged from its golden decade with its theories severely challenged and its principal theorists in profound disagreement.

But the multiverse captured the imaginations of scientists and the broader public alike. Furthermore, it became hugely influential when, around the turn of the twentieth century, string theorists took an interest in the idea. The mathematical wizardry of string theorists imbued Linde's bubbling multiverse with yet another layer of variation, filling it not only with empty island universes and islands full of galaxies, but with islands differing in every other conceivable respect as well. This brings us to the next stage on our journey: Does the multiverse really offer an alternative perspective on cosmic fine-tuning? Can it crack the riddle of design?

FIGURE 32. *Stephen Hawking and Andrei Linde (standing next to Hawking) in Moscow in 1987, with Andrei Sakharov (seated) and Vahe Gurzadyan.*

CHAPTER 5

LOST IN THE MULTIVERSE

Er hat den archimedischen Punkt gefunden, hat ihn aber gegen sich ausge-
nutzt, offenbar hat er ihn nur unter dieser Bedingung finden dürfen.

Man found the Archimedean point but he used it against himself. It
seems that he was permitted to find it only under this condition.

—Franz Kafka, *Paralipomena*

"I HOPE YOU'LL MAKE BLACK HOLES," STEPHEN SAID WITH A BROAD
smile. We exited the cargo lift that had taken us underground into the
five-story cavern housing the ATLAS* experiment at the CERN lab, the
legendary European Organization for Nuclear Research near Geneva.
CERN's director general, Rolf Heuer, shuffled his feet uneasily. This was
2009, and someone had filed a lawsuit in the United States, concerned that
CERN's newly constructed Large Hadron Collider, the LHC, would produce black holes or another form of exotic matter that could destroy
Earth.

The LHC is a ring-shaped particle accelerator that was built, princi-
pally, to create Higgs bosons, the missing link—at the time—in the Standard Model of particle physics. Constructed in a tunnel underneath the
Swiss-French border, its total circumference is twenty-seven kilometers
(almost seventeen miles), and it accelerates protons and antiprotons[1] running in counter rotating beams in its circular vacuum tubes to 99.9999991
percent of the speed of light. At three locations along the ring, the beams
of accelerated particles can be directed into highly energetic collisions,

* ATLAS stands for A Toroidal LHC Apparatus.

re-creating conditions comparable to those reigning in the universe a small fraction of a second after the hot big bang, when the temperature was more than a million billion degrees. The tracks of the spray of particles created in these violent head-on collisions are picked up by millions of sensors stacked like mini–Lego blocks to make up giant detectors, including the ATLAS detector and the Compact Muon Solenoid, or CMS.

The lawsuit was soon to be dismissed on the grounds that "speculative fear of future harm does not constitute an injury in fact sufficient to confer standing." In November of that year the LHC was successfully turned on—after an explosion at an earlier attempt—and the ATLAS and CMS detectors soon found traces of Higgs bosons in the debris of the particle collisions. But, so far, the LHC hasn't made black holes.

Why wasn't it entirely unreasonable though for Stephen—and Heuer too, I think—to hope that it might be possible to produce black holes at the LHC? We usually think of black holes as the collapsed remnants of massive stars. This is too limited a view, however, for anything can become a black hole if squeezed into a sufficiently small volume. Even a single proton–antiproton pair accelerated to nearly the speed of light and smashed together in a powerful particle accelerator would form a black hole if the collision concentrated enough energy into a small enough volume. It would be a tiny black hole, for sure, with a fleeting existence, for it would instantly evaporate through the emission of Hawking radiation.

At the same time, if Stephen and Heuer's hope to produce black holes had come true, it would have signaled the end of particle physicists' decades-old quest to explore nature at ever shorter distances by colliding particles with ever increasing energies. Particle colliders are like microscopes, but gravity appears to set a fundamental limit to their resolution, because it triggers the formation of a black hole whenever we increase the energy too much trying to peek into an ever smaller volume. At that point, adding even more energy would produce a bigger black hole instead of further increasing the collider's magnifying power. Curiously, therefore, gravity and black holes completely reverse the usual thinking in physics that higher energies probe shorter distances. The endpoint of the construction of ever larger accelerators doesn't appear to be a smallest fundamental building block—the ultimate dream of every reductionist—but an emergent macroscopic curved spacetime. Looping short distances back to long distances, gravity makes a mockery of the deeply entrenched idea that the architecture of physical reality is a neat system of nested scales that we can peel off one by one to arrive at a fundamental smallest constituent. Gravity—and therefore spacetime itself—seems to possess an anti-reductionist element, a difficult-to-grasp but important idea to which I return in chapter 7.

So at what microscopic scale does particle physics without gravity transmute into particle physics with gravity? (Or, put differently, how much would it cost to fulfill Stephen's dream of producing black holes?) This is a question that has to do with the unification of all forces, the topic of this chapter. The search for a unified framework that encompassed all basic laws of nature was already Einstein's dream. It bears directly on whether multiverse cosmology really has the potential to offer an alternative perspective on our universe's life-encouraging design. For only an understanding of how all particles and forces fit harmoniously together can yield further insights in the uniqueness—or lack thereof—of the fundamental physical laws, and hence at what level one can expect them to vary across the multiverse.

MOST VISIBLE MATTER is made of atoms that consist of electrons and a tiny nucleus, which itself is a conglomerate of protons and neutrons. Atomic nuclei are held together by the strong nuclear force that acts on quarks, the constituent particles of protons and neutrons. The strong force is strong, but it has an extremely short range, dropping to zero sharply beyond distances of about a ten-trillionth of a centimeter. The second nuclear force, the weak force, acts both on quarks and on a second class of matter particles that includes electrons and neutrinos known collectively as leptons. The weak force is responsible for the transmutation of some nuclear particles into others. For example, an isolated neutron is unstable and will decay after some minutes into a proton and two leptons in a process mediated by the weak nuclear force. The third and final particle force, the electromagnetic force, is the most familiar. Unlike the strong and weak nuclear forces, electromagnetism, like gravity, has a very long range. It operates not only on atomic and molecular scales, binding electrons to atomic nuclei and atoms in molecules, but acts across macroscopic distances as well. So not surprisingly, together with gravity, electromagnetism is re-

sponsible for most everyday phenomena and applications, from communication devices and MRI scanners to rainbows and the northern lights.

All visible matter and the three particle forces that govern its interactions are bundled together in a tight theoretical framework: the Standard
Model of particle physics. Developed in the 1960s and early 1970s, the
Standard Model is a quantum theory that describes matter particles as well
as forces in terms of fields, the undulating substances spread out in space
that we have encountered before. According to the Standard Model, matter
particles like electrons and quarks are nothing but local excitations of extended fields. Particle-like excitations of the force fields that act between
matter particles are known as exchange particles or bosons. Photons, for
example, the exchange particles mediating the electromagnetic force, are
the individual particle-like quanta of the electromagnetic force field.

Its theoretical underpinnings in terms of quantum fields profoundly
shape how the Standard Model conceives the microscopic workings of the
particle world. Take the interaction between two electrons. When two elec-

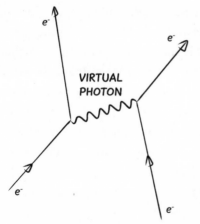

FIGURE 33. *A so-called Feynman diagram depicting the quantum scattering of two electrons through the exchange of a photon. Feynman's sum-over-histories formulation of quantum mechanics stipulates that all possible exchanges must be considered, including those involving more than one photon, to compute the net scattering angle of the electrons.*

trons approach each other they deflect and scatter because like electric
charges repel. The Standard Model describes this process in a tangible manner in terms of the exchange of a photon between the two electrons. When
two electrons enter each other's sphere of influence, it says, one electron
emits a photon and the other absorbs it. As part of this exchange both electrons experience a little kick, which puts them on diverging trajectories (see
figure 33). But that is not all. Feynman's sum-over-histories formulation of

quantum mechanics stipulates that one must add all possible ways one or more photons can be exchanged between the two electrons in order to compute their net scattering angle. This multiplicity of exchange histories means that one can't quite pin down exactly where and when the interaction actually happened, a manifestation of Heisenberg's uncertainty principle.

Now, while photons are massless, just like gravitons conveying gravity, the bosons responsible for the weak and strong nuclear forces are very heavy. This is why the nuclear forces are short-range forces that operate only on the microscopic scales of atomic nuclei. In general, the bigger the mass of the exchange particle, the smaller the range of the force it conveys. It is the masslessness of their microscopic quanta that makes electromagnetism and gravity reach across the universe.

So is this it for the Standard Model? Not quite! There is one last particle, the notoriously elusive Higgs boson, named after the British theoretical physicist Peter Higgs, who postulated its existence in 1964. The Higgs boson is the particle-like quantum of the Higgs field, an invisible scalar field that, much like the inflaton field in the early universe, is thought to permeate all of space, a bit like a modern variant of the ether. The Higgs field is the crucial piece of the Standard Model that gives all other elementary particles their masses. Electrons and quarks, and even the exchange particles, have no intrinsic mass in the Standard Model theory but acquire their masses from the resistance they experience as they move through the all-pervading Higgs field. It is as if particles are constantly wading through the mud when they move about, and the resulting drag is what we call mass. The amount of mass that particles end up with depends on how strongly they feel the Higgs field. Quarks interact very strongly with the Higgs field and are heavy, whereas the lighter electrons do so much more weakly, and photons, which don't interact with it at all, remain massless.

The idea of a scalar field that would endow other particles with mass was first proposed by the timid Higgs and independently by a more flamboyant duo, the American Robert Brout and the Belgian François Englert. The particle-like excitation of the field became known in Belgium as the Brout-Englert-Higgs boson, and as the Higgs boson elsewhere. It forms the capstone of the Standard Model and was finally found with the LHC nearly fifty years later, in 2012, in a discovery that counts as a real triumph of a long and deep symbiosis of curiosity-driven science, advanced engineering, and international cooperation. Much like the discovery of dark energy in cosmology, the experimental discovery of the Brout-Englert-Higgs boson shows once again that empty space is not empty but filled with invisible fields, one of which is responsible for the mass of the matter that makes up almost everything we encounter in daily life. It also demonstrates that nature really does make use of scalar fields as one of the key ingredients it has at its disposal to shape the physical world. As such, the discovery of the Brout-Englert-Higgs boson lends credence to the existence of a similar field that could have driven inflation in the very early universe.

It takes something like the LHC to create Higgs bosons because the Higgs field strongly interacts not only with other particles but also with itself, bestowing its own particle-like quantum with a large mass, m. By Einstein's $E = mc^2$ this means that it requires a great deal of energy, E, to excite the all-pervading Higgs field sufficiently strongly so that it pinches off, ever so briefly, a single, crackling quantum. As a matter of fact, the LHC succeeds in creating Higgs bosons in only about one in ten billion particle collisions. And these Higgs bosons enjoy but the briefest glimpse of existence, decaying almost instantly into a shower of lighter particles. Nevertheless, by carefully scanning through its decay products, particle physicists have been able to deduce some of the Higgs boson's properties, including the fact that it weighs about as much as 130 protons put together. This may sound heavy, but most particle physicists find this incredibly light. In fact, the Higgs boson's mass is a hundred million billionth of what many physicists would consider a natural value.[2] Its value became even more puzzling in 2016 when, despite a major upgrade, the LHC didn't conjure up any of the new elementary particles that theorists had hypothesized to render the tiny Higgs mass somewhat easier to swallow. A lightweight Higgs is important, however, for if the Higgs were much heavier, protons and neutrons would be heavier, too, far too heavy to form atoms. The unbearable lightness of the Higgs is yet another property that makes our universe propitiously fit for life.

Now, the Standard Model doesn't quite predict the values of the masses of particle species, including that of the Higgs. This is because the theory on its own doesn't fix the strength with which each type of particle interacts with the Higgs field. All in all, the model contains about twenty parameters, key numbers such as particle masses and force strengths, whose often surprising values are not predetermined by the theory but must be measured experimentally and inserted into the formulas by hand. Physicists usually refer to these parameters as *constants of nature,* because they appear to be unchanging across much of the observable universe. With these constants in place, the theory offers an extraordinarily successful description of all we know about how visible matter behaves. In fact, by now the Standard Model is by far the best-tested physical theory ever. Some of its predictions have been verified to an accuracy of no fewer than fourteen decimal places!

Yet you might wonder if there isn't a deeper, yet-to-be-discovered principle that determines the parameter values on which the immense successes of the Standard Model rely. The Higgs mass may appear unnaturally small to us, based on what we know, but is its value perhaps implied by higher mathematical truths? Or perhaps the constants aren't really the same constant numbers all across the universe. Perhaps they very slowly evolve, as part of cosmological evolution. Or maybe they change from one cosmic region to another, giving rise to island universes with non-Standard Models of particle physics?

THE MECHANISM BY which the Higgs field gives particles their mass provides the beginning of an answer to these difficult questions. The Higgs' way of generating mass shows namely that the strength of this field isn't a God-given fact but the outcome of a dynamic process that unfolded when the universe began to expand and cool in the aftermath of the hot big bang. Furthermore, this process involves the random breaking of an abstract mathematical symmetry.

It's an all-too-familiar phenomenon that symmetries of physical systems get broken when they cool down. Think of the transition from liquid water to ice when the temperature drops through zero degrees Celsius. Liquid water is the same in all directions: It possesses a rotational symmetry. Ice crystals, on the other hand, have regular geometric structures that break the rotational symmetry of the warmer liquid water. Magnets are another classic example. The magnetic properties of, say, an iron bar drastically change around the critical Curie temperature of 770 degrees Celsius. At temperatures above the Curie point the jiggling magnetic fields of the individual iron atoms don't line up. Under these circumstances the magnetic field outside the bar averages to zero, reflecting the rotational symmetry of the underlying force of electromagnetism. However, when one slowly cools an iron bar below the Curie temperature, domains with a net magnetic field spontaneously form, producing a qualitatively different state characterized by a broken rotational symmetry, with a magnetic North Pole in a particular (random) direction.

This is a general phenomenon. Symmetries of physical systems tend to get broken when the temperature falls, leading to richer structures and more room for complexity. The Higgs field is no exception. The field reacts to temperature in much the way ordinary matter does. In the immediate aftermath of inflation, when the universe was hotter than one hundred million times the temperature at the core of the sun, the Higgs field would have jiggled around wildly and averaged out to a net zero, much like the magnetization of an iron bar above the Curie point. In this net zero Higgs field permeating the newborn universe all particles would have had zero mass—a highly symmetric state of affairs. As the universe expanded and the temperature fell, however, the Higgs field underwent a transition. This transition was triggered at about 10^{-11} seconds into the hot big bang era, when the temperature dropped below a frigid 10^{15} degrees. At this point, the Higgs' thermal jitters lost much of their strength and the field's behavior became dominated by self-interactions instead. The latter are governed by the field's energy curve, the amount of energy it contains for different values of the field. But much like the inflaton field in figure 27, the energy curve of the Higgs field peaks when the field is zero and is lower at a nonzero value of the field. Hence the highly symmetric zero Higgs field suddenly found itself in an unstable state, not unlike a pencil poised on its tip. Figure 34 reminds us that a pencil will quickly trade symmetry for stability by toppling down, randomly choosing a particular direction. Likewise, the zero Higgs field rapidly condensed, everywhere jumping in strength to an energetically favorable state with a nonzero value. It was this symmetry-breaking transition of the Higgs field to a nonzero value that endowed particles with mass, a vital step on the long road toward complexity.

What's more, the reduction in symmetry from the condensing Higgs field is also what triggered the differentiation of the weak and electromagnetic forces. You see, when the Higgs field was zero, not only were the

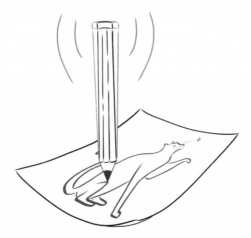

FIGURE 34. *A sharp pencil poised on its tip respects the symmetry of Earth's gravitational field pulling vertically down. However, this symmetric state is unstable, so the pencil will quickly fall down. The pencil's final horizontal state is stable, but it breaks the symmetry of the underlying gravitational field. Unifying particle physics theories predict that in a similar way, the biofriendly laws of particle physics reflect a broken-symmetry state that gradually and randomly congealed when the universe expanded and cooled in the aftermath of the hot big bang.*

matter particles massless, the exchange particles conveying the weak nuclear force were massless too. Sheldon Glashow, Steven Weinberg, and Abdus Salam, the founding fathers of the Standard Model, discovered that in this massless high-temperature situation, physical processes would have been entirely unaffected by particular interchanges of photons with messenger particles mediating the weak nuclear force. That is, the weak force would have been long-range and indistinguishable from the electromagnetic force. There would have been a mathematical symmetry interconnecting both forces, blending them together into a single unified *electroweak force*. But when the temperature of the primeval universe fell through the symmetry-breaking Higgs transition, the unified electroweak force fragmented into the short-range weak nuclear force and the long-range electromagnetic force.

So the picture suggested by the Standard Model of particle physics, applied in the melting pot of the hot big bang, is that the universe wasn't born with the values of the particle masses and force strengths that we have today. Instead these are properties of a broken-symmetry state that only congealed when the universe expanded and cooled. This is a profound insight. It tells us that in the very earliest stages of cosmic expansion some of the basic structure of the physical laws *co-evolved* with the universe they governed. Physicists say that the familiar laws of particle physics are *effective laws*—rules that hold only in the relatively low-energy and low-temperature environment that emerged a little while into the expansion.

It is remarkable that we can discover and use effective laws in particle physics without worrying or even knowing what's happening at shorter distances and higher energies. To this extent nature's hierarchical, nested structure has played out impeccably. You can, for example, describe the macroscopic behavior of water with a hydrodynamic equation that models it as a smooth fluid, glossing over the complicated dynamics of its H_2O molecules. In a similar fashion, you can describe the behavior of a bundle of protons and neutrons at energies below a giga electron volt with a simplified particle theory that ignores that they are made up of trios of quarks. Much of the success of physics in the past has relied on this neat separation of scales. It serves as a warning, really, for the degree of trouble we're getting into when we attempt to include gravity in a unified framework and are confronted with the limitations of this clean nested structure.

Obviously, the exact form of the effective laws produced in this most ancient layer of evolution, hidden deep in the hot big bang, has the most fundamental implications. Imagine if the condensing Higgs field had wound up with a slightly different strength. Then the particle masses would have been different as well. But even modest changes in these would have far-reaching consequences, often precluding the existence of stable atoms, thus compromising chemistry, and, once again, the habitability of the universe.

Within the confines of the Standard Model we can rest assured: The net result of the symmetry-breaking Higgs transition is universal. Yes, the field can slide down from its energy curve in different ways, much like the pencil in figure 34 can topple down in different directions. However, its overall strength and hence the resulting particle masses always wind up being the same. But the Standard Model is only part of the story of particle physics. For one, it puts together the strong and electroweak forces in a tentative manner only. Furthermore, the Standard Model doesn't account for the dark matter that makes up 25 percent of the total mass and energy in today's universe and that may well involve many more species of particles and forces. Finally, the Standard Model leaves out dark energy and gravity, the warping of spacetime.

All this suggests there is ample room for more unifying simplicity and symmetry when we trace the universe's history back to earlier times still. Although we now enter a more speculative realm, it is plausible that the symmetry-breaking mechanism that fragmented the electroweak force in the Standard Model operates more generally, and that even more of the familiar structure in the effective physical laws evaporates as we proceed to higher temperatures and earlier times.

Take the mere existence of matter particles. The observed universe contains about 10^{50} tons of matter but almost no antimatter. This is another of its biofriendly properties, for if the expanding universe had emerged with equal amounts of each, then all particles would have quickly annihilated with their antiparticles, leaving behind a burst of high-energy gamma radiation and no matter at all. Yet when the LHC produces matter in high-energy collisions, it creates exactly the same quantity of antimatter. So how did the universe emerge from its fiery birth with an excess of 10^{50} tons of matter? Something must have broken the symmetry between matter and antimatter in the ultrahot big bang, slightly favoring the creation of particles over antiparticles.

Such hypothetical symmetry-breaking mechanisms, and their associated Higgs-like fields, are part of extensions of the Standard Model that are known as *Grand Unified Theories*, or GUTs, because they combine the electroweak and strong nuclear forces in a grand unifying scheme. In effect, GUTs are pretty much defined by their symmetries. This strategy goes back to Einstein, who in 1905 used a symmetry principle relating space and time as a basis for his special theory of relativity of spacetime. Lorentz grumbled that Einstein was simply assuming what he and others had been trying to deduce, but history was on Einstein's side. Ever since Einstein, abstract mathematical symmetries have become generally accepted as a valid foundation for physical theories.

. . .

IN A COSMOLOGICAL setting, GUTs predict that if we go back to exceedingly high temperatures, many billion times the temperature at the core of the sun, the electroweak and strong nuclear forces would have been essentially one and the same force and there would have been a perfect symmetry between matter and antimatter. But typical GUTs allow for a tiny degree of mingling between the constituent forces in their unifying basket. One consequence of this mingling would be that a positron, the antiparticle of an electron, could turn into a proton—a particle. Even though such transmutations would be exceedingly rare, such mixture would provide a way to create a slight excess of matter over antimatter in a transition that would have broken the primeval GUT symmetry when the universe cooled. In such a scenario all antimatter would subsequently have annihilated itself with matter in the dense primeval gas, inundating the universe with high-energy photons. But there would be a small relic of matter left, no more than one part in a billion, which, almost as an afterthought, would make up the roughly 10^{50} tons of matter of which you and I and everything else on Earth are made. The photons, in turn, would make up the microwave background radiation today, the cold and faint vestige of the biggest annihilation event in cosmic history.

Obviously, Grand Unified Theories are not nearly as well developed as the Standard Model. The daunting energy scales at which the underlying symmetries would become manifest lie far beyond what even the LHC can reach. Neither can we determine which one of the many possible GUTS describes the ultrahot big bang from our scant cosmological observations of that remote era. But if the broad symmetry principles on which they are based prove correct, then we can expect that some of the most fundamental properties of the physical world, like the existence of mass and matter, aren't prior mathematical truths but the result of a series of symmetry-breaking transitions that transformed primordial symmetry into a basis for complexity.

And it doesn't stop there. Even the basic distinction between particles and forces may fade away in the scorching heat of the big bang. In 1974, the physicists Julius Wess and Bruno Zumino conjectured that there may be a very general symmetry indeed, which they dubbed supersymmetry, that interlinks not just different force fields but also force fields with matter fields. If their idea holds water, then even the very distinction between force particles and matter particles may originate in a series of Higgs-like transitions. These would have broken the initial supersymmetry, perhaps generating along the way particles of dark matter governed by additional forces above and beyond the familiar four.

The general trend here is clear: Our best unifying quantum theories of particle physics say that as the universe cooled from the ultrahot big bang, during that first split second, various mathematical symmetries would have been broken, sparking a series of transitions that gradually forged a structured set of low-temperature effective laws. We thus discover a striking, deeper level of evolution, a *meta-evolution* in which the physical laws of evolution themselves change and transmute. Figure 35 sketches this cascade of transitions—some verified, many hypothetical—that would have transformed the universe's pristinely uniform and symmetric beginnings into the differentiated physical environment that would ultimately evolve to become fit for life.

These remarkable insights revive an old idea put forth by Paul Dirac. Back in the 1930s, Dirac was already speculating that the physical laws aren't fixed and immutable truths that are imprinted on the universe at birth like a watermark. "One further point in connection with the new cosmology is worthy of note," Dirac said. "At the beginning of time the laws of Nature were probably very different from what they are now. Thus we should consider the laws of Nature as continually changing with the epoch."[3] Eighty years on, the unifying particle theories, operating in a cosmological setting, embody a realization of Dirac's idea of evolving laws.

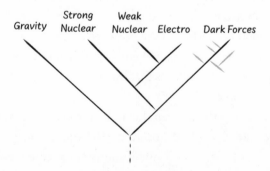

FIGURE 35. *The tree of physical laws grew out of a series of symmetry-breaking transitions in the hot big bang. Unifying particle theories predict that this most ancient layer of evolution could have turned out very differently.*

Furthermore, the random element in their symmetry-breaking transitions lies at the very center of our efforts to understand better why the observed laws are what they are.

For there is another crucial point. The more ambitious Grand Unified Theories that would hold sway at the highest energies don't uniquely determine the outcome of this primeval evolution. Quite on the contrary, the grandest GUTs predict that there are many different ways in which symmetries can be broken, leading to different low-temperature laws by the time the universe ages to a good fraction of a second. This suggests that the properties of the Standard Model and of the dark matter, both of which have had such a decisive impact on the universe's evolution, aren't uniquely determined by the math behind GUTs but reflect, at least partly, the particular outcome of our universe's embryonic history.

This state of affairs is thoroughly familiar from biological evolution. I recalled in chapter 1 that life built its stupendous complexity on an inconceivably large number of frozen accidents throughout history. From the functionalities of individual organisms over the characteristics of species to the taxonomy of the tree of life, law-like patterns in biology encode the outcomes of countless chance events, which, over a period of billions of years, in a co-evolving environment, have enabled layer upon layer of complexity to emerge. Some laws of the living world can even be traced to symmetry-breaking chance events not unlike the cosmic transitions we discussed above. An oft-cited example is the orientation of the helical structure of DNA. All known life-forms on Earth have DNA molecules that wind in a right-handed way. This universality is remarkable, since the laws of electromagnetism on which molecular chemistry relies treat left-handed and right-handed DNA completely equally—symmetrically. Hence, life would have flourished equally well if it were based on left-handed DNA. Although there are various wild hypotheses, it is plausible that around 3.7 billion years ago, when first life was emerging, a random accident caused it to get going with right-handed DNA and that once this symmetry-breaking event had occurred, this particular molecular configuration became part of its basic architecture—a law of life on Earth.

In a similar fashion, GUTs tell us, many of the properties of the effective laws of physics find their roots in accidental twists and turns in the universe's earliest evolution that subsequently got frozen in as part of its physical blueprint. The random component in this ultimately comes about because the laws of particle physics are quantum mechanical, and quantum mechanics is not deterministic. Random quantum jumps of fields in the immediate aftermath of the big bang influence how, exactly, the sequence of symmetry-breakings unfolds. Just as a pencil topples down in a random direction, the exact manner in which the various cosmic transitions have led fields to condense into an amalgam of distinct forces involved an unavoidable element of chance.

On the other hand, not everything is possible. The reason is that the fields in the early universe intertwine with one another; changes in one field affect other fields, and so forth. That interconnectedness, which is ultimately traceable to the fields' common origins, constrains the space of possible pathways. Hence there would have been an interplay of random variation and selection in the earliest stages of the universe's evolution. A sort of Darwinian process of chance versus necessity, playing out at the bottom level of the laws of physics.

The upshot, of course, is that the rules of the cosmic game, the very laws on which the physical universe runs today, might have turned out very differently. There might well have been six species of neutrino instead of three, or four sorts of photon, or a strong interaction between the visible and the dark matter. Such variations yield universes that would be unimaginably different from ours. Grand unification and its even grander super-extensions all point to the stunning conclusion that the relative strengths of the particle forces, the masses and species of particles, and perhaps even the mere existence of matter and forces aren't mathematical truths carved in stone but fossil relics of an ancient and largely hidden epoch of evolution in the wake of cosmogenesis.

Still, you might say, this Darwinian branching of physics happened in (less than) the blink of an eye and in an extremely primitive environment. By contrast, life on Earth evolved over a time span of billions of years in the complex biosphere of this planet, which was itself continuously evolving.

That is true. By a billionth of a second into the post-inflationary expansion, when the universe had cooled to a pleasant billion degrees, the form of the effective laws of physics had basically crystallized. One would have thought this really doesn't leave much room for any kind of Darwinian process to play out. However, what counts when it comes to hammering out effective laws isn't the duration in time but the range of temperatures that the system traverses. The latter is obviously huge in the early universe, leading to numerous transitions and ample room for their cumulative chance outcomes to shape physics and cosmology at lower temperatures.

So how much wiggle room is there? What is the balance between variation and selection when it comes to the fundamental laws of physics? It is well known that in biology, the scope for variation is fantastically large. The number of genes that can be conceived mathematically, let alone their possible sequences in DNA, is far, far larger than any other number we ever encounter. Only a tiny fraction of these molecular combinations is realized in life on Earth. This enormous configuration space in biology means that chance wins—overwhelmingly so—and that biological evolution is a strongly divergent phenomenon. And indeed, the amount of information contained in the tree of life that stems from frozen accidents throughout evolution vastly outweighs what follows from plain chemistry and physics. This is what prompted Gould and others to declare that if we were able to rewind the clock and run biological evolution again, we would end up with a very different tree of life.

But is there an equally wide playing field in the wake of the hot big bang? Is the structure of the branching tree of physical laws depicted in figure 35 primarily dictated by deep mathematical symmetries at its roots, or is it mostly shaped by historical accidents? Evidently this is a crucial point, and one central to the aspirations of multiverse cosmologists.

To get a feel for the array of possibilities, we must take our journey toward a unifying framework one final step further, and include gravity.

AS I ALREADY alluded to, the extension of grand unification to include gravity poses challenges of a wholly different magnitude. To begin with, Einstein's general relativity describes gravity in terms of a rigid classical field—the fabric of spacetime—whereas the Standard Model and GUTs speak of jittering quantum fields. A unified theory, therefore, would seem to require a quantum description of gravity and spacetime. Stephen's Euclidean approach to quantum gravity provides exactly that, at least approximately, but the imaginary-time geometries on which it relies only capture certain general properties of gravity's quantum realm. They hardly elucidate the nature of the microscopic quanta behind spacetime. Even more, working with quantum fields has proven inadequate to obtain a fully fledged quantum description of gravity. This is because quantum jitters of the spacetime field grow stronger without limit on ever smaller scales. Microscopic fluctuations of spacetime create a self-reinforcing cycle of ever more frantic jittering that destroys its own basic structure. And unlike other fields, which undulate in a fixed background of space and time, gravity *is* spacetime. This is the crux of the difficulty when trying to reconcile gravity with quantum theory.

Enter string theory. In the mid-1980s theorists discovered an exciting new track toward quantum gravity by replacing point-like particles with strings as the basic constituents of physical reality. The central tenet of string theory is that if you were to dissect matter on ever finer scales, far smaller than the smallest scales we can reach with the largest particle accelerators, you'd find deeply hidden inside all particles tiny vibrating strands of energy—which physicists have dubbed *strings*.

Strings are to string theory what atoms were to the ancient Greeks: indivisible and invisible. Unlike the Greek notion of atoms, however, all strings in string theory are equal: The exact same type of string lurks inside all species of particles. This egalitarianism certainly fits beautifully with a unifying philosophy, but you may wonder how the same kind of string

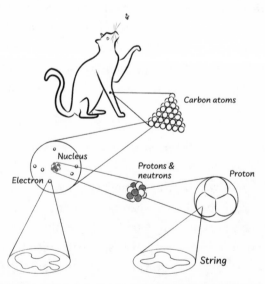

FIGURE 36. *String theory envisions that the microscopic building blocks of matter aren't particles but tiny vibrating strands of energy—strings.*

could possibly endow individual particle species with distinct identities, from their specific masses and spins to their electric charges or colors. The answer, according to string theory, is that a string can wiggle in different ways. String theory asserts that electrons and quarks and even force particles like photons arise from distinct vibrational patterns of a unique type of string. So just as different vibrations of a string on a cello produce different notes, string theory predicts that a universal strand-like entity jiggling in many different ways produces a zoo of different particle species.

Crucially, the pioneers of string theory discovered that in one of its vibrational modes, a string has precisely the right properties to act as a single quantum of gravity—a *graviton*. What's more, by smearing out points into wiggling filaments, string theory tames the troublesome quantum jitters of spacetime on super-small scales. Indeed, there aren't quite super-small scales to begin with in string theory, as the Feynman diagram in figure 37 illustrates. This diagram describes the scattering in string theory of two gravitational quanta. We see that it is impossible to pinpoint the exact location where the two ringing gravitons interact. It is as if the indecomposable string-like building blocks equip the microworld with a minimum length scale below which space is intrinsically fuzzy. This extra layer of uncertainty plays a key role in how string theory prevents the microscopic jitters of spacetime from spiraling out of control.

Remarkably, this extra fuzziness extends even to the shape of spacetime itself. Relativistic spacetime can be warped and curved, for sure, but string theory goes further and says that the geometry of spacetime isn't uniquely fixed, to the extent even that entire dimensions of space can ap-

FIGURE 37. *String theory describes gravitons, individual quanta of gravity, as tiny vibrating loops. This Feynman diagram depicts the interaction of two such string-like gravitons. We see that the process of scattering is smeared out in space and time. This smearing helps control short-range quantum jitters of spacetime.*

pear or disappear. *What is the geometry of spacetime?* we ask in relativity. The answer, according to string theory, is that it depends on your perspective. In string theory there can be different shapes of spacetime that nevertheless describe physically equivalent situations. Such shapes are said to be dual and the mathematical operations linking different geometries are known as *dualities*. The most famous and mind-blowing duality of all, known as the holographic duality, will be the central topic of chapter 7.

By the late 1980s, string theorists were convinced that interacting one-dimensional strings gave a mathematically sound, microscopic description of gravity, and this became the theory's main claim to fame. Before string theory, gravity and quantum theory seemed fundamentally at odds, as if the book of nature had been written in two volumes telling contradictory tales. With the discovery of string theory, theoretical physicists glimpsed how the two central pillars of twentieth-century physics could work harmoniously together after all. Stronger still, both pillars in some sense *emerge* from the unifying string theory framework. When applied to large and massive objects, string theory's rules basically reduce to the Einstein equation of general relativity. When applied to a small number of not too energetically vibrating strings, the same string theory rules yield the usual theory of quantum fields.

Even today, however, the fundamental structure of the theory remains somewhat elusive. In fact, if you were to ask a number of theorists, *What is string theory?*, you are likely to get a range of different answers. In the absence of direct experimental access to the ultrahigh energies at which the string-like nature of matter and gravity would be manifest, string theorists have mostly had to supplant input from experiments with Dirac's dictum "to look for interesting and beautiful mathematics" in order to further develop their theory. It must be said that by and large string theorists haven't been bothered by this. Over the years the string theory community has developed its own intricate checks-and-balances system to judge progress by, based mostly on criteria to do with the mathematical consistency of the framework and the sheer depth of theoretical understanding it offers. And this has produced a remarkably innovative science. By now the field of string theory has evolved far beyond the original goals of merging gravity with quantum mechanics. It has created a web of relations interlinking a broad spectrum of branches of physics and mathematics, from the physics of superconductivity and quantum information theory to, as I will discuss in chapter 7, quantum cosmology.

However, unlike the Einstein equation of general relativity, or the Schrödinger and Dirac equations of quantum theory, a single agreed-upon master equation that encapsulates the kernel of string theory has yet to be found. What's more, the unifying power of string theory has come at a cost, and not a trivial one: For the math behind string theory to work out, the strings must move around in nine dimensions of space (*modulo ambiguitatis*). That is, the rules of string theory require there to be six extra spatial dimensions besides the familiar length, width, and height in order for the theory to hang together mathematically.[4]

You might wonder why these extra dimensions don't immediately rule out the theory as a viable description of our world. Surely we would have noticed if there were more dimensions of space? This isn't necessarily true, however, because the six extra dimensions could be extraordinarily small and curled up tightly at every point, instead of stretching over cosmic scales like the three familiar dimensions. If this were the case it could be very difficult to establish their existence. It would be much like looking at a drinking straw from a great distance. The straw looks one-dimensional, even though it has a second circular dimension going around it that would be visible if one were having a drink holding the straw in one's hand. Likewise, if the six extra dimensions are much smaller than the length scales that the LHC or any other high-energy experiment can currently discern, then their existence will so far have escaped our notice. The six-dimensional spatial blob that might be lurking at every point in space will hitherto have looked like, well, a point (see figure 38).

But because strings are so tiny, they do wander into the hidden six-dimensional world. And just as the shape of a cello determines the combination of vibrational patterns that creates its unique tone color, the geometry of the six-dimensional nugget in string theory determines what particle species and forces the wandering strings breathe into existence.

String theory paints our visible three-dimensional world as a sort of shadowy reflection of a far more complex visible, higher-dimensional reality that we discern only indirectly.

So the picture that emerges is that the nature of matter and the form of the effective physical laws in the three large dimensions of our

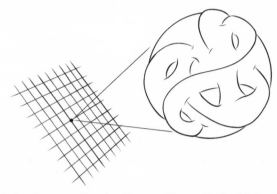

FIGURE 38. *String theory predicts that if we could magnify the fabric of space enormously, we would find that every point in the familiar three large dimensions consists of tiny extra dimensions. Moreover, the shape of this extra-dimensional nugget hiding at every point impacts the amalgam of forces and particles that exist in the three large dimensions.*

experience—including the strength of the particle forces, the number and types of particles, visible and dark alike, their masses and electric charges, and so on, and even the amount of dark energy—all depend on the way six small dimensions lurking at every point are curled up.

But what principle selects the shape of the tiny spatial knot that corresponds to the biofriendly macroworld we observe? That is the riddle of design in the tantalizing mathematical landscape that string theory opens up.

THE FOUNDING FATHERS of string theory held out high hopes that a powerful mathematical principle at the core of the theory would single out *the* unique shape of the extra dimensions. With string theory, it was thought, we were well on our way to explaining the key numbers behind the Standard Model and the pie chart of the universe on the basis of pure mathematical reasoning. Plato would soon be vindicated, the mantra went, and the striking biofriendliness of the universe would turn out to be nothing but a fortuitous consequence of its rigid mathematical underpinnings.

Those early hopes were soon dashed with the discovery that, just like musical instruments, extra dimensions in string theory come in a huge number of shapes and forms. During the 1990s, theorists watched with horror as a stupendous number of new ways to wrap six extra dimensions into a tiny nugget were found. These hidden geometries can be extremely complicated, with a multidimensional labyrinth of geometric handles, bridges, and holes, wrapped and pierced with fluxes of field lines, all folded tightly like origami. In figure 38, I have attempted to evoke such an elaborate shape, although the projection down onto a two-dimensional sheet means the picture really doesn't do justice to the higher dimensional complexity featuring in string theory.

Combining the various ingredients in their theory, string theorists found far more possible shapes of hidden dimensions than there are atoms in the observable universe. Each shape of space composes its own string symphony, describing a universe with its own specific set of effective laws. So by exploring the mathematical landscape of extra dimensions, theorists were also uncovering a mind-boggling cornucopia of effective physical laws in the three large dimensions that we do see. Certain special arrangements of hidden dimensions correspond to universes with laws nearly identical to the ones we observe, differing, say, only in the precise values of a few particle masses. Such universes could be equally biofriendly—if not more. The vast majority of extra-dimensional wrappings yield universes completely unlike ours, with an entirely unfamiliar amalgam of particles and forces. By the turn of the century, string theory had become a Walmart for physical laws; one could dream of a particular universe governed by certain effective physics and then go and find some configuration of extra dimensions matching it. There are countless universes where the repulsive effect of the dark energy prevents the formation of galaxies and life, the odd universe where the LHC would have produced black holes—and Stephen would have gotten his Nobel Prize—and even universes in which a different number of space dimensions expand and grow large.

In a cosmological context, the molding of the extra dimensions is part of the chain of symmetry-breaking transitions that causes the tree of effective laws to bud. Also the burst of inflation, the transition whereby three space dimensions break away and expand, can be viewed as part of that sculpting of the high-dimensional reality in the wake of the universe's birth. In fact, even the birth itself, in which space might have "split" into spacetime, has the flavor of a symmetry-breaking transition, in some sense the ultimate one. Moreover, quantum jumps add an element of randomness to this entire process. And although most of these jumps leave no trace, those triggering symmetry-breaking transitions get amplified and frozen in as part of the newly emerging effective laws. This is the Darwinian interplay of variation and selection again playing out in the primitive environment of the very early universe—the most ancient and lowest level of evolution we can imagine.

The stupendous range of six-dimensional knots means that string theory's answer to the question I posed above is that variation and chance win over necessity. By a great deal. Is the theory of everything so powerful, then, that it determines nothing?

On the one hand, the fact that vibrating quantum strings generate gravity means that string theory has everything it takes to realize Einstein's dream of a complete unified theory of all forces and particles. Moreover, unlike the Standard Model, string theory has no free parameters that must be measured before the framework can be put to use. From a theoretical viewpoint this is as pure as it gets in physics. On the other hand, this purity apparently enables the theory to harbor a breathtaking myriad of effective laws. In his wonderful book *The Hidden Reality,* Brian Greene describes this mind-blowing mathematical landscape in great detail and elaborates on no fewer than five very different ways in which the intricate structures of string theory wiggle a multitude of effective laws into existence.

From a practical viewpoint, this Walmart of laws means that string theory is not a law but a metalaw. In hindsight, this shouldn't come as a surprise, for the absence of parameters and predetermined structures in the unifying math behind string theory implies there must be an emergent element to any effective laws it encodes, and emergence in a quantum world is subject to random variations.

Curiously, therefore, the story of grand unification can be read in two ways, each exhibiting a different side of the story.

Reading from low to high energies, from top to bottom in figure 35, we recover the success story of the unification program of particle physics. Going up in energy we encounter ever more embracing symmetries, encoding ever deeper mathematical patterns that interconnect the observed forces and particles and possibly even the dark matter in an ever more inclusive unifying framework. This is the orthodox, particle physics reading of unification, and it is how these ideas are tested in the lab. Particle physicists request larger accelerators to smash particles together with higher energies in order to explore deeper-lying unifying symmetries (all the while assuming that the threshold to create black holes lies higher still). This reading also emphasizes the interdependencies among nature's building blocks and the kernel of necessity that unification seeks to uncover.

Reading from high to low energies, from bottom to top in figure 35, we see a sequence of transitions that create a branching, treelike structure of physical forces and particle species very much reminiscent of the tree of life (see figure 5). This is the natural reading in a cosmological context, where expansion causes cooling and cooling causes branching. Viewed this way, grand unification is in the first place a grand source of variation, enabling a process whereby physical laws mutate and diversify much like biological species would do billions of years later.

These readings aren't in contradiction with each other. They are merely two sides of the same coin—variation and selection.

THE MOMENTOUS DISCOVERY of string theory's staggering spectrum of branching pathways proved a game changer for the theory of the multiverse. Early on, multiversists like Linde and Vilenkin had already realized that we can expect island universes, regions where inflation ended and transitioned into a hot big bang, to differ in their structure and composition. Some would have enough matter to produce billions of galaxies, while others would be nearly empty. With string theory, the scope for variation among island destinations skyrocketed to unimaginable heights. String theory predicted that if there is an eternally inflating arena out there, then it plays host to a staggering diversity of island universes. Each one would bear the imprint of its birth, with its own cascade of transitions when it expanded and cooled. The multiverse as a whole would be a truly bewildering, variegated cosmic quilt, stitched together somehow by the invisible hand of string theory's metalaws.

Inhabitants of a given island universe looking around might get the impression that the physical laws are universal and—as it happens—they might even wonder whether the laws were carefully crafted to bring forth life. But in the variegated multiverse of string theory this would be an illusion. What we call the "laws of physics" would be local patterns only, frozen relics reflecting the particular way our patch of space cooled from its hot big bang. Much like the pointed beaks of tree finches or the right-handedness of DNA, the properties of particles and forces wouldn't be part of a grand design but merely features of our cosmic environment. It's just that the Darwinian-like process that gave rise to the effective physical laws happened far, far back in the past, hiding their evolutionary character.

I vividly remember Leonard Susskind's lecture "The Anthropic Landscape of String Theory,"[5] at Universe or Multiverse?, one of the first scientific conferences that brought string theorists and cosmologists together. The meeting was held at Stanford University in March 2003 and had been convened by Linde and Paul Davies. There was a jubilant mood among the gathered theorists. For years, progress toward a final theory that uniquely lined up with the observed world had stalled. In a stunning reversal, Susskind at the meeting argued that this pursuit had been misguided. String theory rests on firm and profound mathematical principles, he explained, but the theory isn't a physical law in the usual sense. Instead we should think of it as a metalaw that governs a multiverse of countless island universes, each with its own local laws of physics.

Later that year, down south in Santa Barbara, the Kavli Institute for Theoretical Physics ran its first Superstring Cosmology program. In a packed auditorium, a beaming Linde explained to an audience of string theorists hanging on to his every word how his universe-generating mechanism of eternal inflation could blow into existence an endless proliferation of island universes occupying even the most remote corners of the string theory landscape. Eternal inflation, he propounded, turns the enormous scope for variation in the theory into a real cosmic patchwork—the multiverse.

Worryingly, however, there was nothing in the metalaws of string theory that said where we were supposed to find ourselves in this crazy cosmic quilt and hence what kind of universe we should expect to observe around us. The multiverse on its own is impersonal and incomplete. That is the paradoxical situation I described in chapter 1: As a physical theory, the multiverse signals an end to much of the predictivity gained by physics.

But Susskind proposed a grand new deal. By combining the multiverse with the anthropic principle, he argued, this explanatory embarrassment can be turned around, for the anthropic principle *selects* a biofriendly patch in the multiverse. Taken on its own, the anthropic principle wouldn't qualify as science, but in combination with the multiverse, he advanced, the anthropic principle has real predictive power. So he put forward "anthropic multiverse cosmology" as a new paradigm for fundamental physics and cosmology, to replace the orthodox framework based on objective and timeless laws only.

IN HINDSIGHT, WHAT really triggered the anthropic "revolution" in cosmology was a remarkable interplay between these new theoretical insights in string theory and fresh observations that pointed to an invisible dark energy permeating space. I mentioned earlier that around the turn of the twenty-first century, and to nearly everyone's surprise, astronomical observations of supernovae indicated that the universe's expansion has been speeding up for the past five billion years. Theorists, left scrambling for an explanation, resurrected Einstein's notorious λ term, whose dark energy and negative pressure cause gravity to repel on the very largest scales. The amount of dark energy—the value of the cosmological constant λ— required to account for the observed acceleration is extraordinarily small, however: an astounding 10^{-123} of what many would consider a natural value. This embarrassing discrepancy between what we expect and what we observe has to do with quantum mechanics, which predicts that empty space should be teeming with virtual particles, jitters of the quantum vacuum. The energy associated with all this frenzied activity in the vacuum actually generates a cosmological constant. But when particle physicists add the contributions of all virtual particles, they find an absurdly large number for λ, a cosmological constant of such large magnitude that it would rip the universe apart before galaxies could even begin to form. Now, up to the late 1990s most theorists assumed that there was a yet-to-be-discovered symmetry principle at the heart of string theory that uniquely fixed the dark energy to be zero. But the discovery in the early 2000s that the theory harbors a sprawling multiverse tied in with the awesome observation that the cosmological constant is not zero after all, to spur a dramatic change of perspective on "what's natural" when it comes to λ. The search for a fundamental explanation of zero was swiftly replaced with a belief that the amount of dark energy changed randomly from one island universe to another in a vast and variegated multiverse, and that the anthropic principle selected the very small but nonzero value we observe.

Interestingly enough, the first anthropic considerations in this context actually predate all these theoretical and observational developments of the late 1990s. Already in 1987, at a time when multiverse speculations were largely brushed off as bad metaphysics, Steven Weinberg performed a most remarkable gedankenexperiment in which he reflected on the value of the cosmological constant from an anthropic viewpoint. Weinberg conceived of a hypothetical multiverse and studied which island universes would develop a web of galaxies. He noticed that this condition placed an extremely stringent upper bound on the local value of the cosmological constant. In fact, island universes in which λ is just a tad larger than the value we observe would begin to accelerate millions instead of billions of years after their big bang, leaving no time for matter to aggregate.[6] Now, without galaxies the universe would be a lifeless place. Hence the fact that we exist, Weinberg concluded, naturally leads us to zoom in on those island universes with only the slightest trace of dark energy, lying in an extraordinarily narrow biofriendly window.

On the other hand, we should not expect the dark energy density to be much smaller than what is required for our existence. This was the anthropic touch to his argument. Weinberg assumed that we are randomly selected observers living in a multiverse of island universes that sample nearly all possible values of λ, even within the narrow band of values compatible with life. The vast majority of habitable island universes would have densities of dark energy near the upper bound set by life, simply because it requires excessive fine-tuning to select an even smaller value of λ. Reasoning along these lines, he concluded that the observed amount of dark energy shouldn't be zero, as was generally believed at the time, but in fact as large as it can be, as long as it didn't disrupt the formation of galaxies. This conclusion led Weinberg to predict, as far back as 1987, that astronomical observations might one day reveal that the cosmological constant doesn't vanish but takes on a very small nonzero value. Within a decade, the supernovae observations proved him right.

What's more, string theory appeared to provide the enormously variegated multiverse that Weinberg in his gedankenexperiment had assumed existed. And so it happened that in a remarkable chain of events the triangle of observations, theorizing, and anthropic reasoning on λ—each one revolutionary in its own right—came together in the early 2000s. It is this confluence of ideas that led to anthropic multiverse cosmology, the new paradigm that gave wings to the sweeping change of perspective on cosmic fine-tuning that Susskind and others embraced.

If there is a multiverse out there, then, just by chance, there will be rare island universes here and there possessing local laws that are suited for life. Obviously life will emerge only in those island universes. Other island universes where the conditions aren't biofriendly will remain unobserved, simply because we won't be observing where we can't be. The anthropic principle serves to select the habitable islands in the multiverse, even if these are exceptionally rare. So, taken together, anthropic multiverse cosmology would seem to resolve the age-old riddle of design: We inhabit a rare biofriendly patch picked out by the anthropic principle in a mostly lifeless cosmic mosaic.

At first glance this line of thought doesn't seem very different from the way we account for ordinary selection effects within our observable universe. We cannot exist in regions of the universe where the density of matter is too low for stars to form or in an era before there was a good supply of elements like carbon. Rather, we live on a rocky planet surrounded by an atmosphere in the habitable zone of a specially stable and tranquil stellar system, many billion years after the big bang, because this is a particularly biofriendly environment where intelligent life has had a chance to develop. Likewise, anthropic multiverse cosmology says, we have intrinsically biofriendly laws of physics merely because we could hardly have evolved in a universe where the physical conditions preclude our existence. In a sense the anthropic principle says that we find the physics of the observable universe the way it is because we are here.

Stephen wasn't impressed. He very much agreed with Susskind, Linde, and their followers that the universe's conspicuous biofriendly design demanded an explanation. He profoundly disagreed that anthropic multiverse cosmology explained anything. On our way back to Pasadena after the Santa Barbara meeting, we stopped at a Cuban dance club in Beverly Hills where Stephen took out time from dancing to voice his dissatisfaction with the new string cosmology. *Advocates of eternal inflation and the multiverse get themselves tied in knots, on what a typical observer would see,* he typed out on the rhythms of Cuban salsa. *In their picture we are all Chinese, and disaster is around the corner.*

. . .

BY THE TURN of the century Stephen had grown increasingly concerned that anthropic reasoning in cosmology undermined the rational method that is science's beating heart. This is perhaps a subtle point to understand, but bear with me, for we have arrived at the core of the matter in Linde versus Hawking.

The basic problem is that the anthropic principle relies on the assumption—all too often swept under the carpet—that, in one way or another, we are typical inhabitants of the multiverse. In order to reason anthropically, that is, one must first specify what's typical and what's not. This is done by singling out some of the biophilic properties of the physical world that are deemed important for life. These properties transcribe words like "we," "us," or "the observer" in the language of physics. Their prevalence, together with the statistical properties of the multiverse, is then used to deduce in what sort of island universe we—typical multiverse dwellers— should be, and hence what sort of physics we can expect to discover with our telescopes.

But what selects the "anthropic" properties that specify the ensemble of observers of which we are supposedly typical, randomly selected members? Should we consider certain properties of the effective laws, or count the number of spiral galaxies, or even the fraction of baryons that end up in galaxies, or the number of advanced civilizations? We are all typical in some respects and atypical in other ways. Am I typical in the sense that I live in the most populous country on my planet? Readers from India will answer yes, others won't. Am I typical in the sense that I live in a country that has four seasons? Most readers will answer yes but some won't. Furthermore, there are vast numbers of ensembles for which there is no way of telling. Do we live in the universe that has the largest number of civilizations? We might. But we might just as well live in a universe that is atypical in this respect, in that it has some civilizations but far fewer than other universes. Neither present nor future data will tell. We simply don't know. And this is a problem for anthropic multiverse cosmology. Because in the absence of a clear-cut criterion that specifies *the* correct reference class of multiverse inhabitants, all theoretical predictions of anthropic multiverse cosmology become ambiguous. The theory falls prey to personal preferences and subjectivity. From your viewpoint we should find ourselves in one type of island universe while my perspective selects another island universe, with no possibility for a rational resolution on the basis of evidence from experiment and observation.

Weinberg's anthropic "prediction" that we should measure a small but nonzero cosmological constant provides a case in point. Closer inspection has revealed that the predicted value of λ depends strongly on the chosen reference class of multiverse inhabitants. Weinberg assumed we are typical among inhabitants of island universes that differ in their amount of dark energy but otherwise share exactly the same physical parameters. Cosmologists Max Tegmark and Martin Rees pointed out that if instead we assumed we are randomly selected among the much larger ensemble of observers inhabiting island universes differing both in their cosmological constant *and* in the size of their galactic seeds, then the predicted value of λ would be a thousand times larger than what we measure.[7] More creative choices of reference classes lead to truly absurd conclusions to the point that we should expect to be vacuum-fluctuated brains in an otherwise empty island universe, with all our memories fluctuating into existence a fraction of a second ago. The bottom line is this: Reasoning anthropically, one can always turn any setback into an apparent success, or vice versa, by appropriately adjusting the population of random observers to one's liking.

Now, it may be asking too much of multiverse cosmology to be falsifiable in the old Popperian sense. Nature may not be so kind. There may not exist a smoking gun measurement that would enable us to rule out definitively the entire multiverse theory. However, it is not asking too much of a physical theory to make unambiguous predictions, so that further observations and experiments have at least the potential to strengthen our confidence in it. Without this, the process of science becomes compromised. Like any theory that relies for its predictions on random distributions from which you can observe only one instance—ours—anthropic multiverse cosmology fails to meet this most basic criterion.

Cosmologists refer to this as the *measure problem* of multiverse cosmology: We lack an unambiguous way of measuring the relative weight of different populations of island universes and this undermines the predictive power of the theory when it comes to *our* observations. In fact, this measure problem manifests itself perhaps most sharply when we attempt to predict properties of the universe that have nothing directly to do with life, for then the anthropic principle provides no escape route to hide behind.[8]

We have arrived at a textbook Kuhnian crisis. The hope was that the anthropic principle would specify "who we are" in the cosmic patchwork of eternal inflation, that it would relate abstract multiverse theory to our experiences and measurements as observers within this universe. However, it fails to insert "us" into the equations in a manner that complies with basic scientific practices, leaving the theory with no explanatory reach at all.

As a matter of fact, invoking a process of random selection apart from and on top of the metalaws conceals a decidedly non-anthropic, godlike perspective on cosmic affairs. Random selection imagines that we are somehow looking down on the cosmos from the outside and "choose" who we are among all observers like us. This would be justifiable practice if we or some metaphysical agency had actually performed such an operation and was aware of the result of this selection. However, there's not a shred of evidence for this. It is fallacious reasoning to equate our mere noticing that we are human beings living in a universe to performing a cosmic act of random selection.[9] Hence we should not be deriving theoretical predictions as though we were randomly selected from an ensemble of our choice. Indeed, it is perfectly possible (and not necessarily unlikely) for us to live in a universe with effective laws that are atypical in many respects. In fact, this is just what the randomness of the symmetry-breaking transitions should lead one to expect.

There is the observed universe with its effective laws and its configuration of stars and galaxies, occasionally harboring life. Whether this is all there is or whether this exists as part of a gigantic multiverse, the logical situation is the same: Our universe, the one we observe, exhibits a set of physical properties that are supremely suited to bring forth life. Whatever may or may not be happening in distant, causally disconnected universes should be entirely irrelevant when we attempt to understand the design of this one.

THE PITFALLS OF reasoning on the basis of typicality are all too familiar from other historical sciences, from biological evolution to human history. Had Darwin assumed that we are typical, he would have considered an ensemble of earthlike planets with a variety of trees of life, all containing a *Homo sapiens* branch. He would then have attempted to predict that *we*—this particular instance of *Homo sapiens* on planet Earth—should be part of the most common tree of life among all possible trees with *Homo sapiens* branches. That is, he would have categorically ignored his key insight that each branching involves a game of chance, and that the tree of life as we know it encapsulates the convoluted history of trillions of accidental twists and turns, over billions of years of biological experimentation, rather than an external act of random selection.

The sheer vastness of the possibilities in biological evolution means that any kind of causal deterministic explanation of why we have this particular tree of life is doomed to fail. This is why biologists work *ex post facto,* describing how a given outcome leads back to a specific sequence of branchings. If anything, typicality can be a useful guiding principle to explain a few of the most general structural properties of the biosphere.

String theory conceives of an equally vast space of possible pathways when it comes to forging effective physical laws in the wake of the big bang, a quantum process involving random jumps and a series of symmetry-breaking transitions. As a consequence, a given outcome needs to be neither typical nor a priori likely.[10] Yet unlike modern biology, anthropic multiverse cosmology defies this randomness and clings to a fundamentally deterministic explanatory scheme that puts "why" above "how." The case of biology suggests, though, that this offers a flawed basis for a better understanding of the appearance of design in cosmology. Nobel laureate David Gross, for one, has long held this view: "The more we observe and know about the universe, the worse the anthropic principle fares."[11]

Multiverse theory asserts that there are fundamental limits to the whole idea of evolution. Framing the ancient evolution leading to the effective laws of physics against the fixed backdrop of unchanging meta-laws, multiverse cosmology adheres to what is, after all, a relatively orthodox explanatory scheme in physics. It assumes that all the way down at the bottom of physics and cosmology we will find stable, timeless meta-laws. These metalaws are assumed to take the form of a central master equation governing the cosmic mosaic as a whole, from which the probabilistic predictions for low-energy observations like ours will be computable. Viewed in this grandest scheme, the multiverse is little more than an epicycle on Newtonian epistemology, somewhat like the way the ancients added epicycles onto epicycles in an attempt to rescue the Ptolemaic world model. Evolution and emergence ultimately remain secondary phenomena in multiverse cosmology, somewhat less fundamental. This is the very core of the matter of Hawking versus Linde: whether, deep down, either change or eternity wins.

IS THAT IT? That evening in Beverly Hills, in the aftermath of the propounded anthropic multiverse revolution, with the Cuban band playing in the background, Stephen was ready to ditch the anthropic principle for good. *Let's do this properly,* he said. No longer content to outsource the falsifiability of cosmological theory to a nonscientific principle, we vowed to rethink its basic foundations. The riddle of design was bound to take us deep into the very roots of physics, and we were on our own. String theorists were out in another universe.

FIGURE 39. *Stephen Hawking and the author, in 2006, down in the cavern housing the ATLAS detector at CERN, with ATLAS spokesperson Peter Jenni and ATLAS deputy spokesperson, later CERN director general, Fabiola Gianotti.*

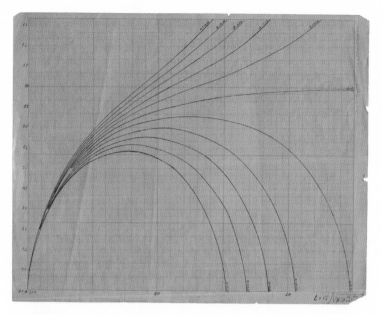

PLATE 1. Georges Lemaître produced this iconic diagram around 1930 showing the evolution of the universe. In the bottom left corner he writes "t = 0," the first moment of time that later became known as the big bang.

PLATE 2. "Who blows up the balloon? What causes the expansion of the universe?" This cartoon depicts the Dutch astronomer Willem de Sitter in the shape of the Greek letter lambda, for Einstein's cosmological constant λ, blowing up the universe like a balloon.

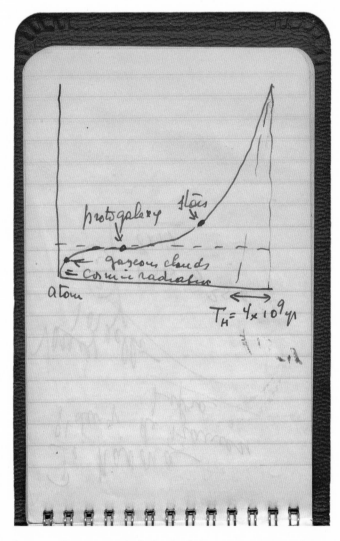

PLATE 3. Georges Lemaître's sketch of a hesitating universe in his purple notebook. Born out of a primeval atom, its wobbly curve of expansion creates physical conditions that make life possible.

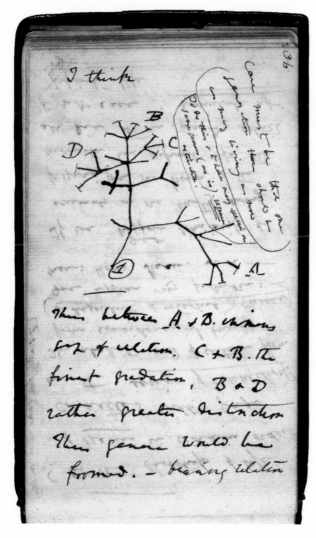

PLATE 4. Charles Darwin's initial tree of life sketch in his Red Notebook B, showing how a genus of related species might originate from a common ancestor.

LEMAITRE FOLLOWS TWO PATHS TO TRUTH

By DUNCAN AIKMAN
PASADENA.

The Famous Physicist, Who Is Also a Priest, Tells Why He Finds No Conflict Between Science and Religion

"THERE is no conflict between religion and science," the Abbé Lemaitre has been telling audiences over and over again in this country and then proving it by explaining the aims of both. His view is interesting and important not because he is a Catholic priest, not because he is one of the leading mathematical physicists of our time, but because he is both. Here is a man who believes firmly in the Bible as a revelation from on high, but who develops a theory of the universe without the slightest regard for the teachings of revealed religion in genesis. And there is no conflict!

Such an attitude would have been preposterous to a Victorian physicist. Either you accept the whole Book of Genesis and therefore shut yourself out of the world of science, or you accept science and repudiate the prophets as expositors of the manner in which the universe began. Today the physicist is meeker. Behind his formulas there is something that is still veiled. He is half mystic and ready to admit that the universe may reveal itself in other ways than in mathematical equations or the bands and lines of a spectrograph. The abbé, therefore, follows the trend of modern thinking and derives from it more than ordinary satisfaction because he happens to be trained in theology as well as in mathematical physics.

Lemaitre, like Eddington, finds that science and religion supplement each other. Science can never study the universe as a whole. It selects a small portion, as much as it can handle, and then makes deductions. To a cosmologist the earth and Mars are only planets wheeling around the sun. Are they inhabited? Are they washed by air and water? Why were they created? Is there purpose in the universe? Science is indifferent to such questions, but not religion.

The questions are just as legitimate as those that are asked by the physicist when he wonders what may be the meaning of a shift to the red in the spectra of distant nebulae. To search thoroughly for the truth involves a searching of souls as well as of spectra. And it is religion that satisfies the soul-searching instinct, according to Lemaitre. In fact, he goes so far as to recommend a course in theology —to him a way of looking at the Bible—to physicists and biologists who see in the Book of Genesis only an interesting piece of ancient folklore.

* * *

LEMAITRE believes that if discussions could be carried on in a friendly, objective way, the church and the laboratory would find themselves closer together than they believe they are. Listen to him as he sits in a student's bare room in the atheneum of the California Institute of Technology, a stoutish young man of 38 who wears horn-rimmed glasses and the expected Roman collar of a secular Catholic priest.

"This conflict," he begins with a smile and a French inflection in his otherwise perfect English—"where is it? Here we have this wonderful, this incessantly interesting and exciting universe. When we try to learn more about it, learn how it began and how it is put together; to find what it is all about, as you say in America, what are we doing? Only seeking the truth. And is not truth-seeking a service to God? Certainly everything in the Bible and in all authoritative Christian doctrine teaches that it is. Has any logical religious thinker of any faith ever denied it?

"Do you know where the heart of the misunderstanding lies?" he asks. "It is really a joke on the scientists. They are a literal-minded lot. Hundreds of professional and amateur scientists actually believe the Bible pretends to teach science. This is a good deal like assuming

that there must be authentic religious dogma in the binomial theorem. Nevertheless a lot of otherwise intelligent and well-educated men do go on believing or at least acting on such a belief. When they find the Bible's scientific references wrong, as they often are, they repudiate it utterly. Should a priest reject relativity because it

contains no authoritative exposition of the doctrine of the Trinity?"

If the Bible does not teach science, among other things, what does it teach? you ask.

"The way to salvation," comes the reply. "Once you realize that the Bible does not purport to be a textbook of science, the old controversy between religion and science vanishes."

"But the Bible says that creation was accomplished in six days," you protest. "Isn't that a direct, literal statement?"

"What of it?" retorts the abbé. "There is no reason to abandon the Bible because we now believe

that it took perhaps ten thousand million years to create what we think is the universe. Genesis is simply trying to teach us that one day in seven should be devoted to rest, worship and reverence—all necessary to salvation."

"And that story about Jonah and the big fish?"

"I admit that a whale cannot

swallow a man and that a whale could not survive the swallowing of a man whole. But what of it? The real lesson is that by faith and righteousness a good man may attain security and salvation whatever his perils may be."

Like Eddington, the abbé believes that some things are imparted to us by revelation. There is no reasoning about the process. There is a lifting of a veil. The means of expressing what is revealed are often faulty, but the truth is there for all that.

So strongly is Lemaitre of this opinion that he is willing to attribute to the prophets all the

powers with which they are credited in the Bible.

"If scientific knowledge were necessary to salvation," he says, "it would have been revealed to the writers of the Scriptures and they would have set it down in their verses. For instance, the doctrine of the Trinity is much more abstruse than anything in relativity or

quantum mechanics. But, being necessary to salvation, the doctrine is stated in the Bible. If the theory of relativity had also been necessary to salvation it would have been revealed to St. Paul or Moses. Even though handicapped by the lack of a terminology and the necessary equations, all the result of an evolution that has been going on for centuries, either would have made some stumbling effort to expound it.

"As a matter of fact neither St. Paul nor Moses had the slightest idea of relativity. The writers of the Bible were illuminated more or less—some more than others—on the

question of salvation. On other questions they were as wise or as ignorant as their generation. Hence it is utterly unimportant that errors of historic and scientific fact should be found in the Bible, especially if errors relate to events that were not directly observed by those who wrote about them. The idea that because they were right in their doctrine of immortality and salvation they must also be right on all other subjects is simply the fallacy of people who have an incomplete understanding of why the Bible was given to us at all."

Lemaitre tells of a classroom scene in which he figured. An old father was expounding at the desk. Before him sat the lad who was to discover the expanding universe and who, even then, was brimful of science. In his eagerness the lad read into a passage of Genesis an anticipation of modern science.

"I pointed it out," says Lemaitre, "but the old Father was skeptical. 'If there is a coincidence,' he decided, 'it is of no importance. Also if you should prove to me that it exists I would consider it unfortunate. It will merely encourage more thoughtless people to imagine that the Bible teaches infallible science, whereas the most we can say is that occasionally one of the prophets made a correct scientific guess.'"

* * *

THERE is, the abbé admits, a varying sense of conflict between science and religion in the different branches of science. "The biologists seem to have peculiar difficulties," he reasons. "There is every reason for this. They have only recently discovered a few guiding laws and principles. Hence, in the past their studies have been confusing rather than enlightening. In a way their subject-matter has been gross.

"But give the biologists more laws like those of the Abbé Mendel and a new spirit is bound to awaken. The sense that this is a morally ordered universe will be inculcated. As soon as any science passes the mere stage of description it becomes a true science. Also it becomes more religious. The mathematicians, the astronomers and the physicists, for example, have been very religious men, with a few exceptions. The deeper they penetrated into the mystery of the universe the deeper was their conviction that the power behind the stars and behind the electrons of atoms is one of law and goodness."

The real cause of conflict between science and religion is to be found in men and not in the Bible or the findings of physicists. "When men were told that they had the right to interpret the Bible's teachings according to their own lights," he holds, "naturally some were bound to decide that its science was infallible and others that it did not agree with modern instrumental measurements and was proof of opposite doctrines. The conflict has always been between those who fail to understand the true scope of either science or religion. For those who understand both, the conflict is simply about descriptions of what goes on in other people's minds."

* * *

AS a priest Lemaitre bows to the Catholic principle of leaving the interpretation of the Bible to the church. But this is good science, too, in his view. "The church has always been aware that the Bible teaches salvation, not science," he insists again. "Although the church's sense of the separate fields of science and religion has unquestionably developed through the ages, its fundamental recognition of the separate but intrinsically harmonious objects of both science and religion has always spared Catholic scientists much confusion."

"And Galileo?" you hint in the hope of tripping him up.

"Oh, Galileo was mildly disci-

(Continued on Page 18)

Einstein and Lemaitre—"They Have a Profound Respect and Admiration for Each Other."

Associated Press Photo.

PLATE 5.

PLATE 6. *The Beginning of the World,* as an abstract, timeless egg by Romanian-born sculptor Constantin Brâncuşi (1920).

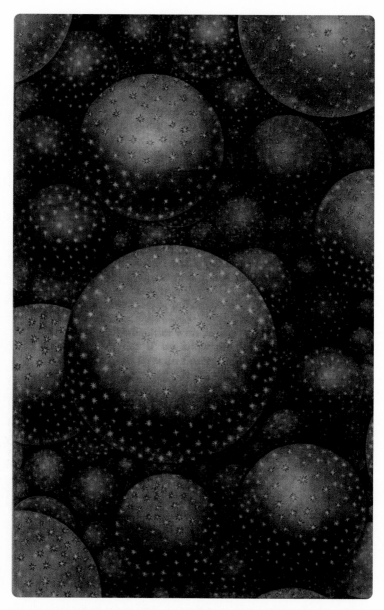

PLATE 7. In *An Original Theory of the Universe* (1750) Thomas Wright imagined a never-ending universe filled with galaxies "creating an endless immensity . . . not unlike the Milky Way." Replacing galaxies with island universes, Wright's picture resembles today's theory of the multiverse, in which new island universes are continually created. What sort of island universe should be ours?

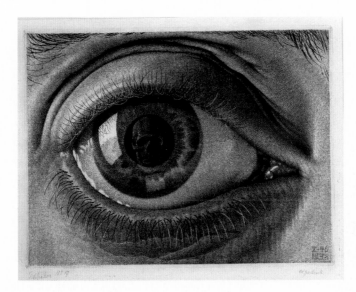

PLATE 8. *Eye* by Maurits Cornelis Escher reminds us of human finitude. We are within the universe, looking up and out, not somehow hovering outside.

$-300 \quad \delta T\ [\mu K] \quad 300$

PLATE 9. The temperature of the relic cosmic microwave background (CMB) radiation as it reaches Earth, at the center of the ball, from different directions in space. The remnant radiation forms a sphere around us that provides a snapshot of the universe a mere 380,000 years after the big bang. This CMB sphere also marks our cosmological horizon: We cannot look farther.

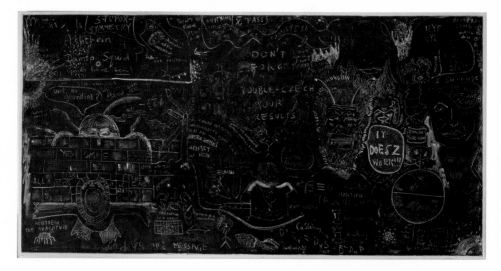

PLATE 10. This blackboard hung in Stephen Hawking's office at the University of Cambridge. It was a memento from a 1980 conference he hosted on supergravity. The early Stephen heralded supergravity as the potential theory of everything.

PLATE 11. Stephen in his Cambridge office in 2012, around the time of his seventieth birthday. In the background, the "second blackboard" features the author's first calculations that would lead us to view the universe as a hologram. The later Stephen held that, in a deeper sense, the theory of the universe and observership are bound together. We create the universe as much as it creates us.

CHAPTER 6

NO QUESTION? NO HISTORY!

We had this old idea, that there was a universe out there, and here is man, the observer, safely protected from the universe by a six-inch slab of plate glass. Now we learn from the quantum world that even to observe so minuscule an object as an electron we have to shatter the plate glass; we have to reach in there. . . .

—JOHN ARCHIBALD WHEELER, *A Question of Physics*

I ONCE ASKED STEPHEN WHAT HE REGARDED AS FAME. *IT'S BEING known by more people than you know,* he replied. I didn't realize quite how modest this reply was until August 2002, when his fame solved a minor cosmic emergency.

This was shortly after my graduation from Cambridge, a few years into our collaboration. My wife and I were traveling to Central Asia along the Silk Roads. I had decided that if I was going to devote the rest of my life to studying the multiverse, I'd better see some more of this universe first. But, as it happened, in Afghanistan, on our way to the great observatory in Samarkand, Uzbekistan, built by the sultan-astronomer Ulugh Beg in the 1420s, I had an email from Stephen urging me to come and see him in Cambridge. Slightly worried, we set off right away. However, trying to leave Afghanistan we got stuck on an old Soviet bridge across the Amu Darya, a river that runs between Uzbekistan and Afghanistan. The lone guard stationed in the middle of the bridge explained that the border crossing was closed to prevent people from entering Afghanistan. I told him we wanted to get out, not in, but that made no difference to him. Back at the Uzbek consulate in Mazar-i-Sharif, trying to negotiate our passage, I showed the kindly Uzbek consul the brief message from Stephen urging me to get back. He was a Stephen Hawking fan, it turned out, and minutes later, he personally drove us across the bridge into Uzbekistan, so we could head off to Cambridge.*

By now DAMTP had moved out of Cambridge city center and become part of a modern campus for mathematical sciences that had been newly constructed behind the playing fields of St. John's College, on the western outskirts of the town. Stephen's spacious, well-lit corner office overlooking the campus and stuffed with—often erratic—domotics couldn't have been more different from the dusty, dark Silver Street office where we had first met. When I rushed in to see him, his eyes were radiant with excitement, and I had an inkling why.

Clicking away a notch or two quicker than usual, Stephen skipped his customary small talk and launched right in.[1]

I have changed my mind. [A] Brief History [of Time] is written from the wrong perspective.

I smiled. *I agree! Have you told your publisher yet?* Stephen looked up, curious.

You took a God's-eye perspective on the universe in A Brief History, I propounded, *as if we are somehow looking at the universe or its wave function from outside of it.*

Stephen lifted his eyebrows, his way of telling me we were on the same wavelength. *So did Newton and Einstein,* he said, as if in defense. He continued. *A God's-eye view is appropriate for laboratory experiments like particle scattering where one prepares an initial state and measures the final state. However, we don't know what the initial state of the universe was and we certainly can't try out different initial states, to see what kinds of universes they produce.*

As we know, laboratories are specifically designed to study the behavior of systems from an external viewpoint. Laboratory scientists meticulously maintain a clean separation between their experiments and the world outside. (And experimental particle physicists at CERN should, indeed, stay safely away from their high-energy collisions!) Orthodox physical theory reflects this separation, with a clear conceptual cut between dynamics, governed by the laws of nature, and boundary condi-

---

* Later we ran into serious trouble trying to leave Uzbekistan because it is illegal to enter via a border crossing that is closed.

tions representing the experimental arrangement and the initial state of the system. The former we seek to discover and test, while the latter we strive to control. That is the dualism I described in chapter 3.

This sharp division between laws and boundary conditions renders laboratory science rigorously predictive but limits its scope, for we can hardly squeeze the entire universe in the straitjacket of a lab. Anticipating Stephen, I responded emphatically: *A God's-eye view is obviously fallacious in cosmology. We are within the universe, not somehow outside it.*

Stephen assented and concentrated intensely on composing his next phrase.

The failure to recognize this, he clicked, *has led us into a blind alley. We need a new philosophy [of physics] for cosmology.*

Ah, I laughed, *time for philosophy at last!*

Leaving his suspicion of philosophy aside for a moment, he nodded—lifting his eyebrows. Linde versus Hawking, we had come to understand, wasn't merely a debate about one cosmological theory versus another. At the heart of the controversy over the multiverse lay central questions about the deeper epistemological nature of physical theory. How do we relate to our physical theories? What do the marvelous discoveries of physics and cosmology ultimately tell us about the great question of existence?

EVER SINCE THE modern scientific revolution, physics has thriven by adopting a godlike view of the cosmos—not as creator—at least not always—but in terms of theoretical perspective.

When Copernicus challenged the ancients' geocentric worldview, he did so by imagining that he looked down on Earth and the solar system from the vantage point of the stars. His assumption that the planets move in circular orbits meant that his heliocentric model wasn't accurate, but neither were the astronomical observations at the time.[2] However, by conceiving Earth and the planets as if he were hovering high above them, Copernicus ushered in a revolutionary new way of thinking about the cosmos and our place in it. He discovered what one might call the Archimedean point in physics and astronomy, the idea that there is a distant viewpoint from which it is possible to leverage an objective understanding.* And while the new science this idea inspired took centuries to develop and change the world, it took but a few decades for the Copernican revolution to open up a whole new conceptual reality in which humankind was no longer the focal point of the cosmos.[3]

Today we know that Copernicus's writings were only the beginning of a relentless pursuit of the Archimedean point. Over the centuries, the Copernican perspective became ever more deeply ingrained in the language of physics. Whatever we do in physics today, whether we accelerate particles, fuse new elements, or capture faint CMB photons, we always reason as if we are dealing with nature from an abstract point outside it—a "view from nowhere" if you want.[4] Without actually being "nowhere," still tied to Earth and its earthly conditions, physicists have devised ever more ingenious ways to act and think about the universe as though we can conceive of it objectively.

No other discovery provided a greater leap forward in this pursuit than Newton's laws of motion and gravity. Newton understood that the relation between mathematics and the physical world, which had mystified scientists since Plato, involves dynamics and evolution, not timeless shapes and forms. The success and universality of his laws reinforced the idea that science was discovering true objective knowledge about the world. Newton attempted to implement a "view from nowhere" in his work by referring all motions to an imaginary fixed stage of space marked out by the distant stars, an absolute space that he took to be unchanging and unmoving. His law of gravity and his three laws of motion dictated how objects move around on that stage, but nothing could ever change absolute space itself. Absolute space, and absolute time, were like a rock-solid scaffolding in Newtonian physics, a God-given fixed and everlasting arena in which everything plays out.

Yet Newton's background of absolutes wasn't quite as objective a point of reference as he had hoped it would be. The simple mathematical form of his laws only holds for those privileged actors on this cosmic stage that don't rotate or accelerate relative to absolute space. For example, suppose

---

* The ancient Greek scientist Archimedes of Syracuse experimented with the lever to lift heavy objects. His feats with the lever reportedly led him to remark, "Give me a place to stand on, and I will move the earth."

you were a "non-privileged astronaut" in, say, a rotating spaceship. If you were to look out the window, you would see the distant stars rotating past in the opposite direction to the spin of your spaceship, even though there are no forces acting on them. This is in violation of Newton's First Law of Motion that says that bodies acted upon by no forces remain at rest, or keep moving at constant speed in a straight line. So, Newton's elegant laws are only true for those special observers who are tied to absolute space, for whom the laws of motion somehow look simpler than they do for everyone else.

This was reason enough for Einstein to be unhappy with Newton's laws. It was anathema to him that we could have a description of nature that privileged certain actors for whom, simply by virtue of their motion, the world looked simpler. To Einstein this was a relic of a pre-Copernican worldview that cried out to be dismantled. And so he did. Replacing Newton's absolute space and time with a new conception of spacetime that is relational and dynamic, his genius was to find a way of formulating physical laws such that all observers would see the same equations at work. The equation of general relativity (see page 42) looks the same for everyone, wherever you are and whichever way you move. In order to account for how the observations of any particular observer depend on their position and motion, the theory comes equipped with a set of transformation rules that relate the perceptions of different observers to one another. These rules allow anyone to extract the "objective kernel" of nature—at least as far as classical gravity is concerned—from this universal equation.

Relativity theory realized Einstein's dream that no one should have a privileged view. For Einstein, the truly objective roots of reality weren't to be found in the particular perspective of a privileged observer but in the abstract mathematical architecture underpinning nature. He moved the quest of physics for an Archimedean point beyond space and time, into the transcendent realm of mathematical relationships. This vision consolidated the idea in scientific circles that there are fundamental laws out there, with an actuality above and beyond the physical universe, supplying true, causal explanations. As the Nobel laureate Sheldon Glashow, perhaps the supreme spokesperson of this position, put it, in 1992, "We believe that the world is knowable. We affirm that there are eternal, objective, extra-historical, socially-neutral, external and universal truths."[5]

Defying the odds, multiverse cosmology has held on to this view that physics ultimately rests on firm and timeless foundations. Multiverse theory in some sense moves the Archimedean point farther out still, much farther than Archimedes, Copernicus, or even Einstein ever dared. Envisioning multiversal metalaws possessing some sort of prior existence, multiverse cosmology reaffirms once again the paradigm, going back to Newton, of a configuration space of physical phenomena, embedded in a fixed background structure, which we can grasp and handle from a god-like perspective.

Now while the ontological status of the physical laws hardly matters in the controlled environment of laboratories, it explodes in our faces when we ponder their deeper origin—let alone when we inquire about their biophilic character. In the previous chapter I have described how multiverse theory gets caught in a self-destructive spiral when one ventures with it into these deeper mysteries. This led us to question whether the entire edifice rests on solid ground. Did the Copernican pendulum in cosmology swing too far toward absolute objectivity?

As a matter of fact, the perplexities of the discoveries of Copernicus and his illustrious contemporaries didn't escape the early modern philosophers. How can we humans, bound to live under Earth's conditions, at the same time be able to see our world objectively? The immediate philosophical reaction to the dawn of the modern scientific age wasn't one of victorious exultation but of profound doubt, beginning with Descartes's *De omnibus dubitandum,* with his far-reaching doubt about whether such a thing as truth or reality exists at all. The momentous insight *Ignoramus,* "We do not know," that sparked the scientific revolution also dealt a blow to human confidence in the world. Hannah Arendt, one of the twentieth century's most celebrated thinkers, sharply articulated this uncomfortable, straddled position in *The Human Condition:* "The great strides of Galileo proved that both the worst fear of human speculation—that our senses might betray us—and its most presumptuous hope—the Archimedean wish for a point outside from which to unlock universal knowledge—could only come true together."[6]

The Cartesian answer to the scientific revolution was to move the Archimedean point inward, to man himself, and to choose the human mind as the ultimate point of reference. The dawn of the modern age threw men back upon themselves. From *Dubito ergo sum,* "I doubt, therefore I am," came *Cogito ergo sum,* "I think, therefore I am." Thus the scientific revolution gave rise to the paradoxical situation in which humankind turned inward while its telescopes, with all the ensuing experimenting and abstraction in their wake, took it outward, millions and eventually billions of light-years into the universe. Five centuries on, the combination of these two opposite movements has left humankind bewildered and confused. At one level, modern science and cosmology have exposed the most wonderful web of interconnections between the nature of the cosmos and our existence in it. From the fusion of carbon in generations of stars, to the quantum seeds of galaxies in the primeval universe, our modern understanding of the cosmos has revealed a marvelous synthesis. At a more fundamental level, however, the level Stephen sought to unlock, these discoveries have left man's grasp of his place in the grand cosmic scheme deeply uncertain. Modern science has created a rift between our understanding of the workings of nature and our human goals, which has infringed on our sense of belonging to this world. Steven Weinberg, ardent reductionist and supremely gifted Archimedean thinker, gave voice to this anxiety at the end of his book *The First Three Minutes,* where he writes, "The more the universe seems comprehensible, the more it also seems pointless."

I cannot help feeling that Weinberg's Platonic conception of the laws lies at the roots of the feeling he expresses here. In a scientific ontology in which we are disconnected from the most fundamental theories of physics and cosmology as such, it is hardly surprising that the universe that science allows us to discover appears pointless, leaving its biophilic character utterly mysterious and confusing.

So what happens if we relinquish a God's-eye view of the world? What if we gave up on the view from nowhere and instead pulled ourselves and everything else down into the system we aim to understand? In a truly holistic cosmology there should be no "rest of the universe" that is kept apart to specify boundary conditions or to maintain a metaphysical background of absolutes. If anything, cosmology is laboratory science inside out—we are within the system, looking up and looking out.

· · ·

TIME TO STOP playing God, Stephen said with a broad smile when we got back from lunch.

The canteen on the new math campus was a far cry from the old DAMTP's bustling common room, which had facilitated so much excellent science and good fellowship. The principal problem with this new canteen wasn't so much that the food was bad but that we weren't allowed to scribble equations on the tables.

For once Stephen seemed to agree with the philosophers. *Our physical theories don't live rent free in a Platonic heaven,* he typed. *We are not angels, who view the universe from the outside. We and our theories are part of the universe we are describing.*

And he continued:

Our theories are never fully decoupled from us.[7]

It's an obvious and seemingly tautological point: Cosmological theorizing had better account for our existence within the universe. The evident fact that we live on a planet in the Milky Way galaxy, surrounded by stars and other galaxies and immersed in the faint glow of the microwave background, means that we necessarily have an "inside-out" perspective on the cosmos. Stephen called this a worm's-eye viewpoint. Could it be that, paradoxical as it might seem, we must learn to live with that subtle element of subjectivity inherent in a worm's-eye view in order to attain a higher level of understanding in cosmology?

While we were pondering these issues, Stephen's office had turned into a dovecote. There was a constant coming and going, from colleagues to medical staff to celebrities, but Stephen seemed oblivious to the frenzy around him. Like so often, I noticed that a good level of chaos was just what he needed to focus his mind. During our customary afternoon break in which he offered me a cup of tea while he devoured a substantial helping of bananas and kiwis, he would zoom in again on the classical underpinnings of multiverse cosmology as the main culprit of persistent godlike thinking in cosmology.

Advocates of the multiverse cling to a God's-eye view, because they assume that, globally, the cosmos has a single history, in the form of a definite spacetime with a well-defined starting point and a unique evolution. This is basically a classical picture.

To be fair, multiverse cosmology is a hybrid of classical and quantum thinking. On the one hand, one imagines random quantum jumps producing a variety of island universes. On the other hand, one assumes this is happening within a giant, preexisting inflating space. The latter serves as a classical backdrop in multiverse theory—a scaffolding not unlike Newton's arena, except that this one would be constantly swelling. This background makes it possible—and tempting indeed—to think about this mosaic of islands as if we are outside, as if the creation of island universes were not fundamentally different from an ordinary laboratory experiment.

Stephen continued clicking to drive home this point. *The multiverse leads to a bottom-up philosophy of cosmology,* he said, *in which one imagines evolving the cosmos forward in time, in order to predict what we should see.*

As an explanatory scheme, multiverse theory subscribes to the ontological programs of Newton and Einstein and their fundamentally causal and deterministic reasoning about the universe. A related manifestation of this thinking is that inhabitants of a given island universe in the multiverse are thought to have a unique and definite past.

But you and Jim conceived your no-boundary theory in the same bottom-up fashion, I submitted, *even though that is supposed to be quantum. That flawed causal view is the vision you laid out in BHT [A Brief History of Time].*

It seemed that with my remark, we had hit upon a crucial point. Stephen raised his eyebrows and swiftly resumed clicking.

While I waited for him to put together a sentence, I browsed through his PhD thesis from 1965 that I found on the shelf behind us. I came across a paragraph toward the end of it where he elaborated on the big bang singularity theorem he'd just proven, stating that it implied that the origin of the universe was a quantum event. Stephen had later gone on to develop the no-boundary hypothesis to describe this quantum origin (see chapter 3). Yet he had interpreted his no-boundary theory through the causal lens characteristic of classical cosmology.

From a bottom-up standpoint, the no-boundary hypothesis describes the creation of the universe from nothing. The theory is regarded as yet another Platonic edifice, as if it were inhabiting the abstract "nothingness" preceding space and time. When Jim and Stephen first put forward their no-boundary genesis, they aspired to give a truly causal explanation for the universe's origin, not just for how it came to be but also why it exists at all. That hadn't gone very well. Viewed as a bottom-up scheme, the no-boundary theory predicts the creation of an empty universe, devoid of galaxies and observers. This, understandably, had made the theory highly controversial, as I described in chapter 4.

Stephen had stopped clicking, and I leaned over his shoulder to read. *I now object to the idea that the universe has a global classical state. We live in a quantum universe so it should be described by a superposition of histories à la Feynman, each with its own probability.*

Stephen was launching into his quantum cosmology mantra. To gauge whether we were still on the same wavelength, I rephrased what I thought he meant: *You're saying we should adopt a full-blown quantum view not just of what's happening within the universe—the wave functions of particles and strings and so forth—but of the cosmos as a whole. That means giving up on the idea there is something like a global classical background spacetime. Instead we should think of the universe as a superposition of many possible spacetimes. So a quantum universe is uncertain even on the very largest scales, on scales well beyond our cosmological horizon like those associated with eternal inflation. And that large-scale cosmic fuzziness puts a bomb under the eternal background that Linde and the multiverse afficionados assume exists.*

To my relief, his eyebrows went up again and clicking resumed, though more slowly this time, as if he were hesitating. But eventually this emerged:

The universe as we observe it is the only reasonable starting point in cosmology.

The oracle level was definitely going up now, intensified by the white steam puffing out of a dehumidifier concealed in an ornament on his desk. Stephen was moving to center stage what philosophers often call the *facticity* of the universe—the fact that it exists and happens to be what it is rather than something else. It sounded reasonable, but where was it leading us? Was he prepared to rethink everything? I had a great many questions, but I had long since learned that whenever Stephen said that something was "reasonable," he meant one or another idea he couldn't quite prove but felt had to be right on intuitive grounds and hence wasn't up for discussion. So I tried to move the conversation on, wondering aloud whether quantum cosmology's more expansive and fluid views of history—from one history to many possible histories—could somehow wean the entire framework of cosmological theory away from the Archimedean point. Could a proper quantum theory of cosmology subsume our worm's-eye perspective within its theoretical scaffold and at the same time, unlike the anthropic principle, hold up basic scientific principles? Five hundred years after Copernicus, that would be a remarkable unification of sorts.

Slowly again, in the midst of the confusion we were running up against in this Kuhnian paradigm shift, and mustering all his energy, Stephen put together one more line:

I think that a proper quantum outlook [onto the universe] will lead to a different philosophy of cosmology in which we work from the top down, backward in time, starting from the surface of our observations. *

I was startled—Stephen's new top-down philosophy would seem to upend the relation between cause and effect in cosmological theory. But when I mentioned this to Stephen, he just smiled. Visibly enjoying the sweet taste of discovery, there was no holding back. On our way out, he drove home our fresh perspective with characteristic concision and ambition:

The history of the universe depends on the question you ask. Good night.

WHAT DID STEPHEN mean? Of course the key role of the "act of observation" in quantum mechanics—the question asked, as Stephen put it—has been recognized since the inception of the theory in the 1920s. It is one of the most surprising features of quantum mechanics that the experimenter's observing and measuring enter explicitly in the process of prediction.

In fact, this feature is what most troubled Einstein about quantum mechanics. When the early generation of quantum physicists reconvened in Brussels in October 1927 for their Fifth Solvay Council, they were celebrating a new, triumphant theory of the microworld. The German physicist Max Born reportedly said that physics would be over in six months, and this wasn't even far from what Ernest Solvay had had in mind in the first place. Solvay had established the councils in 1911 for a period of

---

* By "surface" Stephen meant a three-dimensional slice of four-dimensional spacetime. Strictly speaking, the "surface of our observations" lies just within our past light cone. As an approximation to this, one often considers the three-dimensional spatial universe at a moment of time.

thirty years, because he thought that by then physics would have offered to the world what it has to offer.[8]

For one of the biggest scientific revolutionaries of the twentieth century, however, the new quantum mechanics proved too much to swallow. By the time Solvay V got under way, Einstein had grown deeply uneasy with quantum theory. He had declined Lorentz's invitation to present a paper and was reportedly very quiet during the council. However, the formal meetings were not the only forum for discussion. The scientists were housed in the same hotel, and there, in the dining room, Einstein was much livelier. Nobel laureate Otto Stern left us this firsthand account: "Einstein came down to breakfast and expressed his misgivings about the new quantum theory. Every time he had invented some beautiful experiment from which one saw the theory contains at its core a logical inconsistency. . . . Bohr reflected on it with care and in the evening, at dinner, cleared up the matter in detail."[9]

FIGURE 40. *Niels Bohr and Albert Einstein at the Sixth Solvay Council in Brussels, Belgium.*

Einstein rallied against the quantum idea that a particle could be in a definite place when it was observed and yet only have probabilities to be here or there when it wasn't. "Physics is an attempt to grasp reality as it is, independently of its being observed,"[10] he objected, and he jokingly wondered whether it required a *human* observer for particles to adopt a definite location, or whether a casual glance from a mouse would suffice.

To Einstein, the probabilistic nature of quantum mechanics signaled that the theory was incomplete, that there had to be a deeper-lying framework that permitted an objectively real description of physical reality, regardless of any acts of observation. "The [quantum] theory produces a good deal but hardly brings us closer to the secret of the Old One," he wrote to Born. "I am at all events convinced that He does not play dice."[11]

Niels Bohr, on the other hand, who had a background in philosophy as well as mathematics, had a profound intuition that quantum mechanics was consistent. Bohr took seriously the central tenet of quantum mechanics that observership—the very questions we ask of nature—affects how nature manifests itself. "No phenomenon is a real phenomenon until it is an observed phenomenon," he held.

Solvay V marked the opening salvo of one of the great scientific debates of the twentieth century: Einstein versus Bohr. At stake? The depth and the magnitude of the quantum revolution.

At one level their debate was about the basic status of causality and determinism in physics. With its random jumps and probabilistic predictions, quantum mechanics evidently destroys the direct link, thoroughly familiar from classical physics, between where we are now and where we're going. Is this lack of causality and determinism in our description of nature a temporary expedient—Einstein's position—or a foundational overhaul of physical theory—Bohr's position?

But their debate also takes us into the deeper ontology of quantum mechanics. For in response to Einstein's objections, Bohr was forced to clarify what exactly induces wave functions in quantum mechanics to transition from a fuzzy ghostly superposition of realities into the definite reality of everyday experience. We don't observe a superposition of realities; experimenters find particles either here or there, not both here and there. How exactly does this happen? The bold answer of Bohr's Copenhagen school was that this transition is due to the experimenter's intervention itself. Bohr posited that the act of measuring nudges nature to make up its mind and to reveal either this or that reality. You see, when we decide to measure, say, a particle's position, we must exert an influence on it, for example, by pointing a laser. That influence, Bohr asserted, causes the particle's spread-out wave function to collapse and spike at a single location—the observed location. Let go of the laser and the wave function will spread again, evolving smoothly according to the Schrödinger equation, as I described in chapter 3. Shine and measure, however, and the particle's wave instantaneously coalesces into a state with a definite position.

The trouble with Bohr's scheme was that such sudden collapses are completely at odds with the Schrödinger equation. Wave functions that evolve by the Schrödinger equation don't abruptly collapse but wiggle smoothly and gently at all times. Hence, with his interpretation of what happens in the act of observation, Bohr assigned a special role to observers and their measuring that was radically at odds with the mathematical framework of the theory.

This also meant that the Copenhagen scheme amounted to what is often called an *instrumentalist* interpretation of quantum theory, one that accepts a fundamental discrepancy between what we are capable of measuring with our instruments and the physical reality described by the equations. "Our measurements have as much resemblance to what they really are, as a telephone number has to a subscriber," Eddington once said of the Copenhagen scheme.[12] But such instrumentalism creates a deep epistemological puzzle, for what is quantum mechanics really about, then? The Copenhagen interpretation sheds no light on this conundrum. Indeed it tries to evade this question by predicating a fundamental separation between the quantum world of atoms and subatomic particles governed by the Schrödinger equation, and an external background reality that includes macroscopic experimenters and their equipment, as well as the rest of the universe, all obeying classical laws. The collapse of the wave function in the act of measurement was Bohr's way of bridging these two disjunct worlds, somewhat the way the anthropic principle acts to select an island universe in the multiverse. Both operations were designed to connect an objective mathematical formalism to the physical world of our observations, but they failed because their arching remained extraneous to the basic framework of the theory they were meant to complete.

Bohr and Einstein sharpened their positions over many years and never reached agreement. In hindsight, we value Bohr's profound insight that the process of observation plays a key role in bringing about physical phenomena in a quantum universe. On the other hand, his description of it in terms of the abrupt collapse of the wave function was deeply flawed. Today all the evidence indicates that Schrödinger's math applies not only to microscopic collections of a few particles but also to much larger conglomerates of particles making up macroscopic systems, including laboratories and observers and indeed the universe as a whole. Einstein was right, therefore, not to be convinced by Bohr's scheme. He was wrong, however, to pursue the dream of an alternative theory of physics based on a framework for prediction that once again would render observership irrelevant.

PROGRESS EVENTUALLY CAME from a thorough integration of observership into the mathematical formalism of quantum theory. That synthesis took it far beyond what even Bohr anticipated, and this is where we're heading now.

The road starts with the brilliant work in the mid-1950s of Hugh Everett III, a student of John Wheeler's, who started out working in game theory but then got interested in the quantum measurement problem after listening to a talk by Einstein on the subject. Everett pulled down Bohr's wall separating the quantum microworld from the classical macroworld. His key idea was to take the math behind quantum mechanics seriously and to apply it to everything. Suppose there is no collapse, he suggested, but only a single universal wave function that *includes* observers and everything else, evolving gently and smoothly and, in this process, exploring, à la Feynman, all possible historical pathways. That is, Everett took the monumental step to begin to think about the quantum world in an inside-out manner, as a closed system, without any meddling from outside. Figure 41 evokes this point of view, with Schrödinger's cat together with an observer and his lab all placed in one big box.

Everett's grand challenge, then, was to explain how, say, in measurement situations, the universal wave function can produce a single, concrete answer while at the same time avoiding collapse. This is where his reasoning gets exciting—and shocking.

Everett thought carefully about what actually constitutes a quantum act of observation. When experimenters make a measurement, he reasoned, their interaction with the system they measure entangles first a few particles, then their apparatus, and finally their mental state with the system's quantum state. This entanglement, the Schrödinger equation tells us, causes their combined wave function not to mysteriously collapse (as Bohr argued) but, quite on the contrary, to branch out into distinct wave fragments, one for each of the different possible outcomes of the measurement. Hence by reasoning in terms of a universal wave function that encompasses both observers and observed, Everett was able to keep all possible measurement results in the air. Of course, this also meant that observers, too, split. Observers in quantum mechanics bifurcate into nearly identical copies of themselves—one on each branch—distinguished only by the result of the measurement each recorded.

Take Schrödinger's cat, for example, the famous conundrum described by Schrödinger in which a cat is placed in a sealed box on top of an explosive that will detonate if a radioactive nucleus positioned next to it decays (see figure 41). The probability that this will happen is 50 percent within a given period of time. The laboratory-based Copenhagen interpretation looks at the box from an external perspective and predicts that the cat will be in a zombie-like superposition of dead and alive until the box is opened

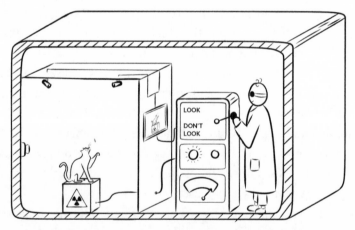

FIGURE 41. *Everett envisaged the universe as a closed quantum system, as a large box containing not only particles and experiments but also observers, their apparatus, and, in principle, everything else. Possible histories for the box universe shown here include whether and when the observer decides to look at the cat, whether the radioactive nucleus has decayed when he looks, how that situation is registered and interpreted in the brain of the observer, etc., etc. Everett sought a formulation of quantum mechanics that predicted probabilities for different histories that describe what goes on in the large box, but without any observation or other meddling with the inside of the box from outside.*

and an observer looks at it, forcing the cat to make up its mind. This does not make sense. A cat can't be semi-dead any more than someone can be semi-pregnant. But Everett's inside-out perspective tells a very different story. It says that in an experiment like this that entangles a cat's fate with that of a radioactive nucleus, the history of the universe constantly bifurcates. In one history, the nucleus decays at a given moment, the explosive detonates, and the cat dies. In the other history, the nucleus does not decay and the cat happily lives on for a while longer. The entire branching process happens smoothly. Neither copy of the cat experiences an unusual superposition, although one copy is of course much better off than the other.

For all practical purposes, then, the individual fragments of Everett's wave function behave like separate branches of reality. Each wave fragment describes a particular historical pathway, consisting of a measurement device registering a particular result, the observer's awareness of it, and everything else that comes along—the laboratory, planet Earth, the solar system, and the large-scale universe. To the observers living in a given branch, the entire bifurcation process proceeds seamlessly, like a river dividing into two streams. None of the observers would be aware of their copycats since they would live out the rest of their lives in different histories, surfing separate crests of the universal quantum wave. "Only the totality of these observer states, with their diverse knowledge, contains complete information," Everett declared.[13]

Everett himself said he sought somehow to bridge the positions of Einstein and Bohr. He claimed their differences were a matter of perspective and described his scheme as "objectively deterministic, with probability appearing at the subjective level." This is an interesting point. In the early Copenhagen formulation of quantum mechanics, probabilities were axiomatic and fundamental. Open a textbook on quantum mechanics from the 1930s, and you will find on one of the first pages that probabilities are *defined* as squares of amplitudes of wave functions. This is not the case in Everett's framework, in which probabilities wiggle their way into quantum theory in a more subtle "subjective" way, much like the way probability enters our thinking in day-to-day life. Whether we ponder the weather, the lottery, or the shape of the next gravitational wave passing through planet Earth, we all use subjective probabilities all the time to quantify our uncertainty in situations where we have incomplete knowledge. This notion of probability was formalized by the Italian mathematician Bruno de Finetti, who in 1974 wrote a treatise stating, "My thesis, paradoxically, and a little provocatively, is simply this: [axiomatic] probability does not exist . . . only subjective probabilities exist, the degree of belief in the occurrence of an event attributed by a given person at a given instant and with a given set of information."[14] And this is what happens in daily life. Throughout our lives, most of us gain confidence in subjective probabilities because we find that results we deem likely happen often, and those that we don't happen rarely.

Deviating from the textbooks, Everett advanced the idea that, just like all other probabilities we use, probabilities in quantum theory are subjective. They arise in his scheme because the ignorance of experimenters regarding which particular outcome they will witness is a source of incomplete information. Probabilities quantify this uncertainty and thus serve as instructions for experimenters to bet on which outcome they will find, much like we use the weather forecast to judge whether we will need an umbrella. The beauty and the utility of quantum theory are that the Schrödinger equation can be used to predict in advance the relative heights of wave fragments that correspond to all possible outcomes of a measurement, and that the squares of these wave amplitudes turn out to be the optimal strategy for placing bets.

AT THE LEVEL of experience, then, every act of observing amounts to some sort of pruning of the branching tree of possible futures. A measurement situation in quantum theory is like a fork in the road, where history divides into two or more separate branches. In the experience of any given observer, at such branching points, only one of the branches survives. More precisely, on each branch, only that branch survives. The branches that do not correspond to the outcome of an observer's measurement evolve independently and are of no further relevance, along with all parts of the tree growing out of them. In a sense they drift away, into the vast unfathomable space of possibilities. Physicists say that such non-interfering branches of history decouple, or decohere.

Not all individual histories decohere, however, a famous example being the interfering trajectories in the double-slit experiment that I discussed in chapter 3. In that setup, electron paths through one slit in the partition do not decouple from those through the other slit but intermingle, producing an interference pattern on the screen (see figure 20). Their intermingling means that we can't tell from observations at the screen through which slit the electron came. It is as if each individual pathway doesn't quite have a separate identity. Only the sum of all interfering paths arriving at a given location on the screen constitutes an independent branch of reality, with a proper probability, and this is how Feynman's sum-over-histories scheme explains the observed interference pattern.

But imagine now a variation of the experiment in which one adds a gas of interacting particles near the slits (see figure 42). When the electron now whisks through the partition, the two wave fragments emerging from each slit will interact with the gas and become rapidly disparate, so that it becomes virtually impossible for them to interfere down the road.

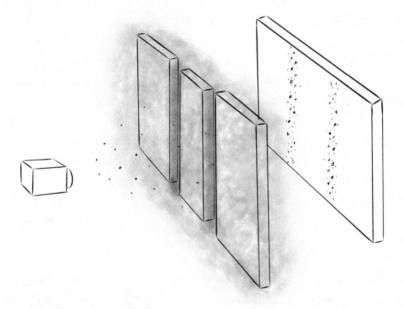

FIGURE 42. *A variant of the double-slit experiment with a gas of particles near the slits that interact with the electrons. Even if the interactions do not affect the trajectories of the electrons very much, they still carry away the subtle correlations among all possible pathways to the screen. As a consequence, the interference pattern is destroyed and replaced with two bright stripes roughly aligned with the two slits, corresponding to the two principal pathways to the screen. The particles effectively perform an act of observation in the quantum sense.*

Not surprisingly, then, the interference pattern on the screen disappears and is replaced with two bright stripes roughly aligned with the two slits, reflecting the two main pathways to the screen. In Everett's language, we say that the environment of particles near the slits has performed an act of observation that causes the wave fragments to decohere into two clearly demarcated histories—branches of reality—that evolve independently from then onward. One might say that the gas of particles in effect asks, *Which slit did the electron take?*, and that by posing this question, it nudges the wave function of the electron to split into two disjunct fragments, corresponding to the two possible answers.

These two variations of the double-slit experiment illustrate two key properties of Everett's scheme. First, the exact nature of the questions we ask impacts the treelike structure of independent branches we have in our bag. Second, meaningful predictions in the form of sensible bets whose probabilities add to one can be made only about properly independent, decoherent historical pathways that differ substantially. We will return to this point in chapter 7, where I discuss what's left of the multiverse once one adopts a quantum view of cosmology.

In the macroscopic world, processes bringing about decoherence are omnipresent. At every moment our environment performs countless acts of observation, washing out quantum interference and transforming a myriad of potentials into a select few actualities. In this way the environment functions as a natural bridge between the ghostly microworld of superpositions and the definite macroworld of everyday experience. Even more, environmental decoherence processes are what make a fairly robust classical reality possible at all, despite the constant quantum jittering on microscopic scales.

Take a high-energy particle released by a radioactive atom like uranium in Earth's crust. At first this particle exists as a wave function, spreading in every possible direction, not quite real until it interacts with, say, a piece of quartz. When that happens, one of its many possible trajectories condenses. The interaction with the quartz transforms what might have happened into what did happen when the uranium atom decayed. Within any given branch of history, this process shows up as a frozen accident in the form of an array of atoms affected by the high-energy particle, the tracks of which are sometimes used to date minerals. The universe we see around us—this branch of reality—is the collective result of innumerable such environmental acts of observation. Having registered and built upon countless chance outcomes, over a period of billions of years, each contributing a few bits of information to our branch of history, this is how the world around us has acquired its specificity. It should not come as a surprise, then, that Stephen, in our conversation, reasoned that a quantum outlook on the universe would slip some sort of backward-in-time element into cosmology.

Mathematically speaking, Everett's scheme is supremely elegant: The Schrödinger equation rules. Universally. Everett's framework demonstrates that Bohr's classical packaging is excess baggage that can be dispensed with. The interactive process whereby subsystems get entangled, causing the universal wave function to divide into separate, decoherent branches that are mutually invisible, offers an extremely satisfying microscopic description of quantum measurements. Human consciousness, human experimenters, and human observations are neither completely irrelevant in Everett's scheme nor regarded as separate external entities, obeying different rules. They are simply treated as part of the broader quantum mechanical environment, not fundamentally different from air molecules and photons. Everett put forward an inside-out way of thinking about the quantum world. He showed that we can be riding the universal quantum wave, not just watching from the shoreline.

THIS ISN'T MERELY a matter of semantics or interpretation. Everett's and Bohr's schemes make genuinely different predictions about how quantum measurements and observations unfold. Whereas Bohr argued that all but one outcome survives, Everett claimed this is only the view from within a given branch of history. His scheme says that to any given observer it appears *as if* the other outcomes have vanished. In Everett's framework, if one were somehow able to reverse all interactions that make up an observation, one could, in principle, recombine the different branches and have them interfere again. Of course, the stratospheric number of particles involved in any act of observation would make this a daunting task in practice. However this would obviously be impossible, even theoretically, if the wave function had collapsed upon observation.

Bohr versus Everett becomes of the utmost importance when we consider the past. You see, Bohr's collapse model prohibits one to even think about retrodicting the past. It is of no use to run the Schrödinger equation backward in time to find out what the past was like, according to Bohr, because countless past acts of observation have interfered with the smooth evolution prescribed by the equation. But retrodicting the past in order to understand how the present came about is central to cosmology. The Copenhagen formulation, therefore, is utterly inadequate for cosmology. It requires the Everettian integration of observership within the mathematical formalism of the theory to make quantum cosmology possible. Everett's scheme brings to the fore a deeper set of principles underpinning quantum theory, principles that have proven crucial to pave the way for its application to the universe as a whole.

At the time, however, Everett's proposal fell on deaf ears. His colleagues either didn't understand what he meant or remained unmoved. The whole idea of applying quantum theory to the entire universe seemed outlandish anyway. Even visionary Wheeler—never shy of grand speculation—felt compelled to add a note to Everett's paper[15] in which he explained his student's formulation of quantum mechanics in a toned-down language that he hoped would make it palatable. All this to no avail. Discouraged and frustrated, and likening his colleagues to the anti-Copernicans in Galileo's day, Everett left the academic world for a military research career.

Much of the community's skepticism arose from the fact that, as a physical picture of the world, Everett's formulation of quantum theory feels bewildering and extravagant. Do we really need an inconceivably large number of unobservable pathways and copies of ourselves just to explain what we observe? It didn't help that Everett's scheme became known as the many-worlds interpretation of quantum mechanics, worlds that are often described as being all equally real, whereas what's really meant is that physical systems have many possible histories.

In the end, however, there was no way around it. Everett's concept of a universal wave function proved *the* foundational insight that made it possible to begin to think about the universe as a whole in quantum terms, as a system *an sich,* neither replicated nor contained in an even bigger box. Everett's work made it possible to hope that a proper quantum outlook onto the universe really had the potential to dispense with a God's-eye view, and to build cosmology anew from a worm's-eye perspective. As such it planted the seeds for the quantum cosmology that Stephen, his Cambridge group, and many others would go on to develop.

THE ARCHITECTURE OF quantum cosmology that grew out of these efforts is sketched in figure 43. It takes the form of an interconnected triptych that includes besides a model of cosmogenesis—the no-boundary hypothesis, for example—and a notion of evolution—Feynman's idea of many possible histories in the landscape of string theory, for example—a key third element: observership.

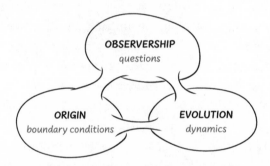

FIGURE 43. *The usual framework for prediction in physics assumes a fundamental distinction between the laws of evolution, boundary conditions, and observations or measurements. For most scientific questions, this split framework suffices. But the riddle of design in cosmology runs deeper, for it inquires about the origin of the laws and our place in the grand cosmic scheme. It necessitates a more general predictive framework that entwines these three entities. A quantum outlook on cosmology provides exactly that. The interconnected triptych sketched here constitutes the conceptual core of a new quantum theory of the cosmos in which evolution, boundary conditions, and observership are folded into a single, holistic scheme of prediction. Its connectedness signals that any laws in quantum cosmology emerge from a mixture of all three components.*

I hasten to say that observership in this scheme doesn't refer to you looking around while riding your bike. Observership in quantum cosmology rather encapsulates the more fundamental quantum act of observation that I have been discussing throughout this chapter: the process whereby at branching points in history one particular result from a range of possible results is converted into a fact. While this process always involves an interaction of some sort, it is by no means restricted to human observations, and the facts generated need not have anything to do with life as such. An observation could be made by a dedicated detector, by Schrödinger's cat, by a piece of quartz, by the breaking of symmetries in the early universe, or even by a lone microwave background photon.

The triptych in figure 43 summarizes the conceptual core of the new cosmology that Stephen and I developed. It envisions that physical reality comes about in a two-step process. First one conceives of all possible expansion histories of the universe, each one originating, say, in a no-boundary beginning. Histories branch out—every branching involving a game of chance—to produce branch effective physics and possibly higher levels of complexity. But this unfathomable realm of uncertainty and potential describes the cosmos only in some sort of preexistence state. At this level there are no predictions, no unifying equation, no global notion of time, no definite anything indeed—only a spectrum of possibilities. Second, however, there is the interactive process we call observership, that transforms some of what might be into what does happen.

Think of Tom Riddle's blank diary in the Harry Potter books. The same goes for the cosmos. The realm of what's possible contains the answers to an endless variety of questions, but it only tells us something about the world by what's being asked of it. In a quantum universe—our universe—a tangible physical reality emerges from a wide horizon of possibilities by means of a continual process of questioning and observing.

WHERE THE FUTURE is concerned, observership is the pruning of the tree of possible pathways stretching out before us. In this process, in the experience of a given observer, only one branch survives. This is Everett's inside-out description of a quantum measurement situation that I described. But observership also reaches into the past. When the Hawkingian oracle said that "the history of the universe depends on the question you ask," I figured this is exactly what it meant. Stephen was saying that the entire collection of facts that characterize the universe around us, from the biosphere on Earth to the observed low-temperature effective laws of physics, constitutes in effect a grand question we ask of the cosmos. The triptych evokes the idea that this grand question retroactively draws into

existence those few branches of cosmological history that have the properties that are being observed. That is, observership in quantum cosmology isn't a mere afterthought or an anthropic post-selection principle
acting in a giant preexisting multiverse, but an agency operating at a
deeper level, an indispensable part of the continual process through
which physical reality—and physical theory, we argue—come about. A
quantum universe and observers emerge in sync, in a sense. The depth of
the top-down philosophy that Stephen anticipated back in 2002—even
though it took us many more years of thought experiments, dead alleys,
and the occasional eureka moments before the mist cleared—is that cosmological theory and observership are bound together.

As I just alluded to, this entangling imbues quantum cosmology with a
subtle backward-in-time element. One doesn't follow the universe from
the bottom up—forward in time—because one no longer presumes the
universe has an objective observer-independent history, with a definite
starting point and evolution. Quite the contrary, built into the triptych is
the counterintuitive idea that in some fundamental sense that I have yet
to elaborate on, history at the very deepest level emerges backward in
time. It is as if a constant flux of quantum acts of observation retroactively
carves out the outcome of the big bang, from the number of dimensions
that grow large to the types of forces and particles that arise. This renders
the past contingent on the present, a further reduction of causality well
beyond what even Bohr conceived.

We are, of course, all too familiar with backward-in-time reasoning
from thinking about other layers of evolution, from biological evolution
to human history. I briefly described in chapter 1 how history at all levels
is shaped by the chance outcomes of countless branching events. These
frozen accidents add a retrospective component to the study of history,
since the vast amount of information they collectively contain is simply
not present in the lower-level laws. It can be gathered *ex post facto* only, by
experimenting and observing.

I recalled in chapter 1 how Darwinian evolution ingeniously integrates
causal explanations with retrospective reasoning into a single coherent
scheme. I venture to claim that, likewise, with the top-down approach to
cosmology, encapsulated in the interconnected triptych in figure 43, we
have found the sweet spot between the why and the how in cosmology. As

we will see, the triptych scheme of prediction is general and flexible enough to bring the deeper questions to do with the riddle of design within its fold.

That being said, the retroactive character of quantum cosmology runs far deeper than the retrospective character of biological evolution. Biologists don't speak of multiple trees of life coexisting in a ghostly superposition until they find fossil evidence favoring one or the other. Instead, they assume, rightly, that we have been part of a given tree of life all along and that we just don't know which one until we have pieced the evidence together. This difference stems from the fact that one can safely ignore the underlying quantum layer in biological evolution. At every branching point in Darwinian evolution, different possible evolutionary pathways immediately decouple from one another because the interacting environment in which life develops instantaneously washes away all quantum interference. That is, the environment continuously transforms, bit by bit, a superposition of trees of life into clearly separated evolutionary trees— one of which is ours. Indeed it takes but a fraction of a second for a gene mutation sparked by a quantum event to decohere. Our tree of life, therefore, has evolved independently from alternative trees long before biologists decide to unearth fossils in an attempt to reconstruct the tree to which they belong. The physical environment has already performed the more fundamental quantum observation. This is not to say, of course, that our becoming aware of the tree of life is irrelevant since, unlike the environment, biologists can interpret their findings and perhaps even use this knowledge to influence future branchings.

By contrast, quantum cosmology inquires about the very origin of the physical environment. It descends all the way to the level of the quantum observation and, not only that, it strives to do so in the remote realm of the big bang where observership has a hand in how the laws of physics came about. Far from being irrelevant, the intermingling in the ghostly world of superpositions is crucial here. It elevates backward-in-time reasoning from a mere retrospective element in the study of this history to a retroactive component that *creates* this history.

It is at this deeper quantum level that the strands interlinking the key ingredients of the triptych become crucially important and that the scheme as a whole takes us well beyond orthodox physics.

. . .

IN THE LATE 1970s, John Wheeler came up with a wonderful gedanken-experiment that did much to clarify this curious element of backward causation in a quantum universe. Wheeler's thought experiment evidenced how the act of observation in the ordinary quantum mechanics of particles can subtly reach into the past, even the remote past.

Wheeler, mentor to both Feynman and Everett, worked with Bohr on nuclear fission before joining the Manhattan Project during World War II. At Princeton University in the 1950s, he revitalized the study of general relativity, taking up where Einstein had left off. With only one precise observational test—the perihelion shift of the planet Mercury—and two qualitative tests—the expansion of the universe and the deflection of light—general relativity had become a bit of a backwater of physics at the time, often regarded as a branch of mathematics, and not even a very interesting one. But, as Wheeler put it, relativity is too important to leave to mathematicians, so he set out to revitalize the field. Wheeler taught the first course on relativity at Princeton, and it included the most privileged annual outing a physics class could ever dream of: a visit to Albert Einstein, at his home on Mercer Street, for tea and discussion.

FIGURE 44. *John Wheeler at Princeton in 1967 lecturing on the differences between classical and quantum mechanics.*

Like Stephen, Wheeler seems to have had a boundless scientific optimism. His imaginative vision and his capacity to bring the biggest questions in physics more sharply into focus inspired lines of research for decades to come. When he passed away in 2008, aged ninety-seven, the *New York Times* obituary quoted Freeman Dyson, saying: "The poetic Wheeler is a prophet standing like Moses on the top of Mount Pisgah, looking out over the promised land that his people will one day inherit."

In his gedankenexperiment on the role of observership and causality in quantum theory, Wheeler considered particles, not universes, because particles are easier to handle. His thought experiment is known today as the *delayed-choice experiment*. It is a variation of a double-slit experiment with photons that was first done by the English polymath Thomas Young in the eighteenth century. In the modern version of Young's experiment, light shines through two parallel slits cut in a partition and hits a photographic plate behind the slits. This produces an interference pattern of bright and dark stripes on the plate, because the distance that the light waves have to travel from either slit to a given point on the screen is in general different. The quantum nature of light becomes manifest when one drastically dims the light source, reducing the waves to a meager stream of photons, emitted one by one. Much like in the experiment with electrons that I described in chapter 3, the arrival of each individual photon particle shows up as a little spot on the plate. But if one runs the experiment for a while in this extremely low-intensity mode, the collection of photon impacts starts to produce an interference pattern. Quantum mechanics predicts this result because it describes each individual photon as a propagating wave function that fragments at the slits, spreads, and intermingles with itself on the far side, creating a pattern of high and low probabilities for where on the plate each photon will land.

If the experimenter decides to "cheat," however, by adding a pair of detectors near the slits that track whether the photons take one path or the other—or both—then the interference pattern no longer emerges. Instead, the photon spots collectively trace out two bright stripes on the plate, the signature of two clearly distinct, classical-ish paths—through one slit or the other. This is because much like the environment of particles in the setup shown in figure 42, placing detectors near the slits amounts to performing an act of observation that causes the wave fragments exiting both slits to decouple. The detectors, by asking which slit the photons whisk through, in effect force the photon wave functions to reveal the particle-like nature of light.

Now, Wheeler envisaged an ingenious variation of Young's experiment in which the detectors aren't placed near the slits but farther down near the photographic plate (see figure 45). In fact, he imagined replacing the plate with a venetian blind and positioning the pair of detectors behind it, each pointing at one of the slits. If we close the blind, the experiment functions as before: The wave function fragments intermingle and produce an interference pattern. By contrast, if we open the blind, the photons simply zip through the plate and the detectors can be used to verify which slit they exited from. In this way the experimenter can decide for each photon individually in which modus to operate the experiment— that is, which question to ask—and thus whether to reveal its particle or wavelike nature.

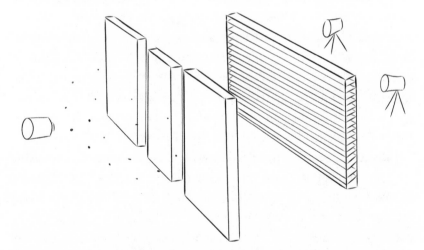

FIGURE 45. *A variant of Young's double-slit experiment with light particles in which the photographic plate on the right is converted into a venetian blind and a pair of detectors are stationed behind it, each pointing at one of the slits. The experimenter operating the detectors can delay his decision right up to the moment each individual photon reaches the blind, whether to leave the blind closed and perform the usual double-slit experiment, producing interference fringes, or open it and verify which slit the photon came through. One might have thought this delayed choice would confuse the photon. Not at all: Nature is smart, the photons always get it right, demonstrating that the act of observation in quantum theory subtly reaches into the past.*

Wheeler's crucial insight was that one can *delay the choice* on whether to open or close the blind right up to the moment the photon gets to the plate. This is an intriguing situation. How do the photons know upon arrival at the partition whether to act as a wave and travel both paths, or as a particle and travel only one, depending on the experimenter's future choice? Quite clearly, the photons can't know in advance whether the experimenter will later open or close the blind. On the other hand, they can't delay their decision whether to be wave or particle either, because if the photon is to be prepared for the possibility that the blind will be closed, its wave function better split at the partition so that the combination of both fragments can produce the observed interference pattern. This seems risky, however, for if the blind turns out to be open after all, because the experimenter happened to decide at the last moment he wanted to know the photon's path, then the wavelike interfering photon would appear to be in trouble.

Actually, Wheeler's thought experiment has since been carried out. In 1984 experimental quantum physicists at the University of Maryland used a high-tech venetian blind, in the form of an ultrafast electronic switch built into a photographic plate, to change between the two modi of operation. They were able to confirm the essence of Wheeler's idea: Photons that hit the "venetian blind" produce an interference pattern; those that are let through do not. Somehow the photons always get it right, even if the choice to turn the path-tracking detectors on or off is delayed until after a given photon has passed through the partition.

How come? Because the unobserved past in quantum mechanics exists as a spectrum of possibilities only—a wave function. Much like electrons or radioactive decay particles, fuzzy photon wave functions morph into a definite reality only when the future to which they give rise has been fully settled, i.e., observed. The delayed-choice experiment illustrates vividly and strikingly that the process of observation in quantum mechanics introduces a subtle form of teleology into physics, a backward-in-time component. The sort of experiments and observations we do today—the very questions we ask of nature—retroactively transform what might have happened into what did happen and, in doing so, take part in delineating what can be said about the past.

Wheeler, forever the optimist, even speculated about a large-scale version of his delayed-choice experiment (see figure 46). He imagined the

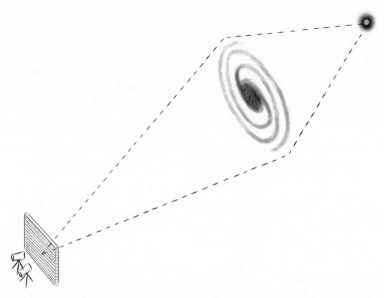

FIGURE 46. *A cosmic-scale version of the delayed-choice variant of the double-slit experiment. The gravitational lensing of a galaxy bends light from a distant quasar. This creates multiple pathways for the light to reach Earth, reproducing the setup of a double (or even triple) slit experiment.*

gravitational bending of light from a faraway quasar by the mass of an intervening galaxy, which then directs it toward Earth. Numerous examples of such gravitationally induced lenses have been spotted in the sky and are routinely used by astronomers to learn more about the amount of dark matter and dark energy in the universe. The deflection means that photons from the quasar can reach Earth via more than one path, by going around the intervening galaxy in different ways, mimicking the situation in a double—or multiple—slit experiment. If astronomers could perform the delayed-choice experiment in this cosmic setting, Wheeler mused, they would be shaping the past billions of years ago, reaching into an era before the solar system had even formed. "We are inescapably involved in bringing about that which appears to be happening," Wheeler wrote.[16]

We are not only spectators.
We are participators.
In some strange sense, this is a participatory universe.

And then he made this remarkable drawing, shown in figure 47, depicting the evolution of the universe as a U-shaped object, with an eye on one side gazing at its own past on the other side, to say that in a quantum universe, observations today impart tangible reality to the universe "back then."[17]

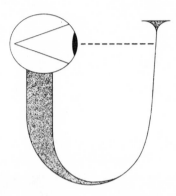

FIGURE 47. *Wheeler thought of a quantum universe as a kind of self-excited circuit. Starting small in the upper right corner, the universe grows in time and eventually gives rise to observers, whose acts of observing bestow tangible reality upon the past, even the distant past when no observers existed.*

Far-fetched in his time, Wheeler's vision of a participatory universe would wind up as a centerpiece of our *top-down cosmology* forty years later. Hawking took Wheeler's observer's participancy seriously—very seriously—and applied it not just to retroactively determine the paths of quantum particles but of the universe as a whole.

The triptych in figure 43 integrates observership with dynamics and conditions into a novel conceptual framework for cosmology. Such a synthesis isn't a mere footnote, or a minor correction to an equation, but a foundational generalization of physics itself. With its unification of dynamics and boundary conditions, the triptych departs from the dualism that has dominated modern physics since its inception. With its inclusion of observership, it gives up on the pursuit of the view from nowhere.

What the top-down character of quantum cosmology does *not* mean, however, is that we can send signals back in time. Observership draws the past more firmly into existence, but it doesn't transmit information back in time. In Wheeler's cosmic-scale version of the delayed-choice experiment, turning our telescopes on or off in the twenty-first century doesn't affect the motion of photons billions of years ago. Quantum cosmology doesn't deny that the past has happened; rather it refines what it means "to happen" and, especially, what can—and what cannot—be said about the past.

Wheeler was fond of illustrating his vision with a variant of the twenty questions game. In this game, a group of colleagues are seated in a living room after dinner. One is sent out. In his absence, the rest decide to play the game with a twist: They agree not to settle on a definite word but to act as if they had agreed upon a word. When the questioner returns and poses his "yes/no" questions, each respondent answers as he pleases, with the one condition that his response should be compatible with all previous ones. So at each stage of the game, everyone in the room has in mind a word that is consistent with all the answers that have been given before. Naturally, successive questions rapidly narrow down the options until both the questioner and the respondents are taken by the hand, as it were, and guided toward a single word. What that final word is, however, depends on the questions the questioner asks and even on the order of the questions. In this variant of the game, Wheeler said, "No word is a word, until that word is promoted to reality by the choice of questions asked and answers given."[18]

In a similar way, a quantum universe constantly puts itself together, piece by piece, out of a haze of possibilities, somewhat like a forest emerging out of the fog on a damp gray morning. Its history is not how we usually think of history, as a sequence of one thing happening after another. Rather it is a marvelous synthesis that includes us and in which what unfolds now retroactively shapes what was back then. This top-down element gives observers, in the quantum sense, a subtle creative role in cosmic affairs. It imbues cosmology with a delicate subjective touch. We—in our observership—are quite literally involved in the making of cosmic history.

"No question, no answer!" Wheeler said of quantum particles. "No question, no history!" Hawking said of the quantum universe.

THE SECOND DEVELOPMENT stage of top-down cosmology, the term Stephen came to prefer,[19] ran from 2006 to 2012. During this period, he developed a deep intuition that with observers as agents integrated within a framework for prediction, we were finally on our way to a cosmological theory capable of elucidating the riddle of design. If only we could understand what, exactly, the triptych was trying to tell us.

Remember that the bottom-up strategy to grasp the biophilic nature of the universe goes as follows: Start with a nugget of space at the origin of time; apply the everlasting objective laws (or metalaws) of physics; watch the universe (multiverse) evolve; and hope it comes out something like the one we live in. This is the orthodox way of reasoning in physics, commonly employed from laboratory experiments and to classical cosmology. A reasoning of this kind seeks a fundamentally causal explanation for the universe's biofriendliness, based on some sort of law-like structure of absolutes. The first bottom-up attempt to untangle the riddle of design was to search for a profound mathematical truth at the kernel of existence. The second line of attack, multiverse cosmology, also relied on timeless metalaws, but augmented with the anthropic selection of a habitable island universe.

But top-down cosmology turns the riddle of design upside down. For a start, it mixes the ingredients in a very different order. The recipe we extract from the triptych reads more like this: Look around you; identify as many law-like patterns in your data as you can; use these to construct histories of the universe ending up like the one you observe; add them together to create your past. So, instead of a background of absolutes, top-down cosmology prioritizes the historical nature of everything. The theory traces the fitness for life ultimately to the fact that deep down at the quantum level, a tangible universe and observership are tied together. The anthropic principle is rendered obsolete in top-down cosmology because the very framework steers clear of the rift, characteristic of bottom-up thinking, that separates our theories of the universe from our worm's-eye view within it. Herein lies the utility of top-down cosmology, and, Stephen felt, its revolutionary potential.

AND SO, WITH the triptych in place, we ran with it. *What shall we do with top-down today?* Stephen would often half-jokingly ask me in the morning.

Now, to get to the core of the early quantum phase of the universe, we must work our way back through the many levels of complexity that separate us from the beginning. This can be done by tracing the evolution of the universe back in time. At first we lose the human and multicellular layers of life and the few law-like rules these obey. Then we lose primitive life, and eventually also the lower-lying geological, astrophysical, and even chemical layers. Finally, we reach the hot big bang era, where the evolutionary character of the physical laws comes to the fore. This was the realm Stephen wanted to venture into.

Let's put the surface of observation all the way back near the end of inflation, he said, *a mere fraction of a second into the expansion. Let's look back from there.*

Equipped with the top-down triptych, ready like a theoretical microscope powerful enough to dissect this lowest level, Stephen was gearing up for the most ambitious thought experiment ever. With a plethora of possible pathways in its basket, quantum cosmology in some sense unpacks the classical big bang singularity. What pops up is a breathtaking deeper level of evolution, taking us *into* the big bang. At this level we discern some sort of meta-evolution, a stage in which the familiar laws of evolution themselves evolve. As I described in chapter 5, the Darwinian-like branching process of variation and selection this entails can only be understood in retrospect. This truly ancient layer of evolution, must be conceived from the top down—looking backward in time.

Take the number of large dimensions of space. According to string theory, the realm of possibilities contains histories with every possible number of large dimensions, from 0 to 10. No prior reason has ever been found why precisely three dimensions grew large and the rest did not. Hence a bottom-up philosophy cannot explain why our universe should have three large dimensions. A top-down approach, however, tells us this isn't quite the right question. Top-down cosmology retrodicts that the observation performed by the most primitive environment in the earliest stages of expansion that three dimensions broke loose and began to inflate carves out those few histories among all possible ones that end up with three large dimensions. The probability distribution over dimensions is of no significance because "we" have already measured that we live in a universe with three large dimensions of space. It would be like asking for the probability of the tree of life compared to that of completely different trees, including ones without a *Homo sapiens* branch. This is neither relevant nor computable. As long as the realm of possible expansion histories contains some universes in which three dimensions expand, it doesn't matter how rare these are compared to histories with any other number of large dimensions. Moreover, this is regardless of whether three is the only number suited for life. Having rendered the anthropic principle obsolete, top-down cosmology treats biofriendly properties on equal footing with everything else indeed.[20]

The same goes for the Standard Model of particle physics. According to Grand Unification and string theory, the Standard Model with its twenty or so seemingly fine-tuned parameters is far from the unique outcome of the sequence of symmetry-breaking transitions in the hot big bang. In fact, there is mounting evidence that paths ending up with the Standard Model are extremely rare in the string theory realm, just as the tree of life on Earth is presumably extraordinarily rare among all possible trees. Once again, therefore, a causal bottom-up approach fails to explain why the universe should end up with the Standard Model. The top-down paradigm approaches this question very differently. It envisages that observations "performed" in the early universe—the outcomes of which are encoded in the frozen accidents that make up the effective laws—carve those histories consistent with the standard model out of the vast spectrum of all possible cosmological histories.

But perhaps the most striking implication of a top-down view of cosmogenesis has to do with the strength of the primeval burst of inflation. Remember that as a bottom-up theory, the no-boundary hypothesis predicted the absolute minimum amount of inflation, barely enough for a universe to exist. By far the most prominent branches of the no-boundary wave function are nearly empty universes that emerge with just a wisp of inflation (see figure 31). That is, if we ignore that we are sentient beings made of atoms and moving about in spacetime and for a moment take a God's-eye view instead, looking at the shape of the no-boundary wave *as if* we are not part of it, then we find we shouldn't exist. This state of affairs had been Stephen's biggest headache in cosmology for more than two decades. The no-boundary hypothesis had the ring of truth to Stephen—very much so—and yet it seemed wrong.

Enter top-down. Adopting a worm's-eye view, top-down cosmology reasons inside out and backward in time. And what happens? The shape of the no-boundary wave dramatically changes. A top-down approach relegates the wave fragments corresponding to empty universes to the far tail of the wave and amplifies those universes that are born with a strong surge of inflation. I illustrate this in figure 48. A comparison with the bottom-up view of the no-boundary wave in figure 31 shows that, quite literally, top-down cosmology completely reshuffles the branches making up the wave function. Furthermore, since the heights of different wave fragments specify their relative likelihood, this means that top-down cosmology retrodicts that the universe started out with a major burst of inflation, in line with our observations.[21] Evidently Stephen was pleased with how this worked out. *At last,* he told me, adding, as if I hadn't realized, that *I have always had a good feeling about the no-boundary proposal.*

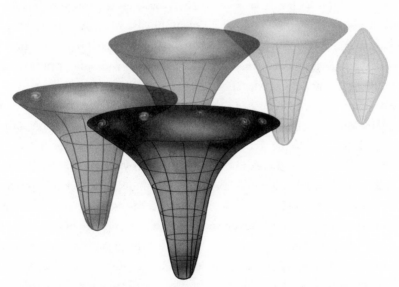

FIGURE 48. *The shape of the no-boundary wave from a top-down perspective. When looked at from the top down, the no-boundary hypothesis retrodicts that our universe came into being with a major burst of inflation, giving rise to a web of galaxies, in line with our observations. The nearly empty universes that dominated the bottom-up wave (see figure 31) recede into the distance.*

. . .

THIS REMARKABLE TURNAROUND in the fortunes of the no-boundary hypothesis provides a vivid illustration that deep down, the past is contingent on the present. But then what, exactly, is the role of a theory of the origin if we view the universe from the top down anyway? One might say that the no-boundary hypothesis is to cosmology what LUCA, the last universal common ancestor, is to biological evolution. Clearly, the biochemical composition of LUCA doesn't determine the tree of life that will grow out of it. On the other hand, there can be no tree of life without LUCA. Likewise, the no-boundary origin is crucial to the existence of the universe, but it doesn't predict the particular tree of physical laws that will emerge from so simple a beginning.[22] Instead, a detailed understanding of the genealogy of the cosmos and its laws can be gathered from observations only—from the top down.

Put differently, models of the origin is a crucial source of predictivity at a more fundamental level. Working top-down, the bowl-like beginnings featuring in figure 48 function as key anchor points for the countless possible pathways veering off into our past. Quantum cosmology without a theory of the beginning would be like CERN without accelerated particles, chemistry without a table of elements, or the tree of life without a trunk. There would be no predictions whatsoever. Any treelike structure that has evolved with interconnected branches rests, ultimately, on the idea of a common origin. Modelling that origin is a key part of any scientific description of the tree. This applies to the tree of life as much as it does to the tree of laws. I venture to claim that there can be no genuine Darwinian revolution in cosmology without a notion of a real beginning. Indeed the lack of a proper theory of initial conditions in, say, multiverse cosmology may well be the fundamental reason why the theory has failed to predict anything.

Nonetheless, you might wonder what we hope to gain from molding a past based on our collective cosmological observations, which leads, obviously, back to what we observe. If top-down cosmology doesn't seek a causal explanation for why the universe and its effective laws are what they are, if it doesn't *predict* that the universe had to turn out the way it did, then wherein, exactly, lies its utility?

Much like Darwinian evolution, the utility of the theory lies in its capacity to unlock the interconnectedness of the cosmos. The theory enables one to identify novel correlations between what may seem at first sight independent properties of the universe. Think about the temperature variations in the CMB radiation. The statistical features of these line up near-perfectly with those of fluctuations generated in universes with a strong inflationary burst. Reasoning from the top down, these are by far the most probable universes. Hence the top-down theory predicts a strong correlation between the observed variations in the CMB, and other parts of our data that select a significant surge of inflation in the first place. Through correlations of this kind, and through predictions of correlations between current and future data, top-down cosmology has great potential to uncover the hidden coherence encoded in the universe. That is why this theory works much better than the multiverse theory with its paradoxical loss of predictivity.[23]

As a picture of physical reality, too, the top-down universe differs radically from the multiverse. In multiverse cosmology, the giant inflating space, bubbling with a myriad of island universes, is just out there (see insert, plate 7). The cosmic quilt exists regardless of which islands have life or which ones are observed. Observers and their observing wriggle their way into the theory as a post-selection effect, without affecting in any way the large-scale structure of the cosmos.

In Stephen's quantum universe, on the contrary, observership is at the center of the action. The top-down triptych restores the subtle thread linking the observer and the observed. Any kind of tangible past in top-down cosmology is always an observer's past. It is as if quantum cosmology conceives of observership as the operational headquarters in the unfathomable realm of all that might be. I have attempted to evoke this "worldview" in figure 49 with yet another branching treelike structure. We are acting and observing (in the quantum sense) and, in this process, growing roots that carve out possible pasts, as well as a select few branches outlining possible futures. The fact that all roots in figure 49 connect on to our observational situation—and this includes what we know about the effective laws—means that the complexity of this treelike structure pales in comparison with that of the multiverse. The vast majority of island universes bears no resemblance whatsoever to the universe we observe.

FIGURE 49. *The quantum universe. Observations today grow roots of possible pasts and outline branches of possible futures, out of the vast realm of "what could be."*

Hence roots corresponding to these don't appear in the quantum tree. Those island histories are gone—lost in an ocean of uncertainty.

I must emphasize, however, if need be, that top-down cosmology remains a hypothesis. We are in a position not unlike Darwin's in the nineteenth century, with data far too sparse to reconstruct in any detail how the tree of laws emerged in the hot big bang. Our fossil evidence of that remote era remains patchy. Take the dark matter, or the dark energy, which together comprise 95 percent of the universe's content. What is the cascade of symmetry-breaking transitions that led to the forces and particles governing the dark sector? Only time will tell.

Given such limited evidence, there remain ardent pre-Darwinians among my fellow cosmologists who hold firmly to a bottom-up view of the world. They maintain that the task of cosmology is to find a truly causal explanation for the universe's judicious design. In their philosophy, chance and historical accidents—let alone observership—must take a back seat. They presume that one way or another the universe *had to* turn out this way on the basis of sound everlasting principles. The top-down philosophy challenges this premise in its ontological essence, with chance and necessity—frozen accidents and law-like patterns—treated on equal footing. If anything, we predict that future observations will reveal many more accidental twists and turns.

YET WHEN I reflect on the long road to top-down cosmology, it becomes clear that we weren't all that much driven by philosophical considerations. (How could we have been, with Stephen on the team.) Rather, we were searching for a better scientific understanding, motivated by a desire to resolve the paradoxes of the multiverse and to crack the riddle of design. As a matter of fact, after Jim and Stephen put forward their no-boundary hypothesis in 1983, they parted ways. Stephen felt we understood quantum mechanics well enough and saw no further need to sift through its foundations. "When I hear the words 'Schrödinger's cat,' I reach for my gun," he once said, and he went on to try to put his no-boundary proposal to the test. But Jim wasn't so sure we understood quantum mechanics well enough, so he turned away from quantum cosmology. Working with the late Murray Gell-Mann, the erudite polymath who in 1964 postulated the existence of quarks, Jim went on to develop Everett's quantum ideas for particles and matter fields. Their foundational work, combined with that of many other physicists,[24] ultimately led to a fully fledged new formulation of quantum theory, known as *decoherent histories quantum mechanics*. This formulation much clarified the physical nature of the branching process in Everett's scheme and, crucially, has observership firmly embedded within its conceptual scheme.[25] So, when in 2006, I realized that Jim and Stephen's insights would have to be merged if quantum cosmology were to live up to its potential, I brought them together again, and it is this inspired move that heralded the second stage of the development of our top-down approach.

Truth be told, however, I believe that the top-down triptych is roughly how Lemaître and Dirac, in their poetic pioneering of quantum cosmology, envisioned it would eventually play out. In 1958, at the Eleventh Solvay Council, on the Structure and Evolution of the Universe, Lemaître gave a status report on the primeval atom hypothesis.[26] After noting that "the splitting of the Atom can have occurred in many different ways"—Everett's branching!—and that "there would be little interest to know their relative probabilities"—no typicality!—he continued: "Deductive cosmology cannot begin before the splitting has proceeded far enough to reach practical macroscopic determinism"—in other words, our expanding branch must decohere in order for a bottom-up approach to be viable. Lemaître concluded his report with this cryptic remark: "Any information on the state of matter at this moment [just after the splitting of the atom] must be inferred from the condition that the actual universe has been able to evolve from it"—an early glimpse of a top-down viewpoint indeed.

This being said, with the exception of these few cryptic comments, top-down cosmology finds its most tangible grounding in Wheeler's prophetic thought experiments and his vision of a participatory universe.

In a recent tribute to Wheeler,[27] Kip Thorne recalled a lunch with him and Feynman in 1971 at the Burger Continental near Caltech, a diner Stephen too frequented while at Caltech.

Over Armenian food, Wheeler described to us his idea that the laws of physics are mutable. "Those laws must have come into being. . . . What principles determine which laws emerge in our universe?" he asked. Feynman, Wheeler's student in the 1940s, turned to Thorne, Wheeler's student in the 1960s, and said, "This guy sounds crazy. What people of your generation don't know is that he has *always* sounded crazy. But when I was his student, I discovered that if you take one of his crazy ideas and you unwrap the layers of craziness from it one after another like lifting the layers off an onion, at the heart of the idea you will often find a powerful kernel of truth."

When Stephen and I embarked on our top-down approach to cosmology, I wasn't aware of Wheeler's ideas, though I suspect Stephen knew about them at least vaguely. In hindsight, we learned that we were unwrapping quite a few layers of Wheeler's craziness, transforming his grand intuition into a proper scientific hypothesis.

. . .

WE DROVE OVER to Gonville and Caius College, Stephen's college and his other base in Cambridge. It was Thursday and that meant dinner in the college, followed by the Fellows' quaint rituals over cheese and port in their paneled Combination Room. With port orbiting clockwise around the long wooden table and the fire crackling, we chatted about the Silk Roads. Stephen reminisced about his travels to Iran in the summer of 1962, to Esfahan and Persepolis, the capital of the ancient Persian kings, and across the desert to Mashhad in the east. *I got caught in the Buin Zahra earthquake* (a massive 7.1 Richter quake that caused more than twelve thousand fatalities), he told us, *when I was on the bus between Tehran and Tabriz on my way home. Still, I would like to go back,* he added. *There should be no boundaries to scientific collaboration.*

While the other College Fellows retired to their rooms and Stephen's nurse nudged us to get going too, he instead settled in for a late-night discussion. I wasn't surprised. Directing his attention to his software program Equalizer again, he set out to talk. I walked around the table to take a seat next to him.

I wrote in Brief History . . .

I completed the thought for him: *that we are just a chemical scum on a medium-sized planet orbiting an average star in an ordinary galaxy.*

His eyebrows went up in assent.

That was the old bottom-up Hawking appeared on the screen. *From a God's-eye view we are but an irrelevant speck.*

Stephen turned his eyes in my direction, reflecting, I think, on the distance he had traveled since BHT. There you have it, I thought, his farewell to the worldview in which he had invested so much.

About time for a change of worldview? I ventured. The chimes of the college chapel's bell rang across the courtyard. Stephen hesitated again. I decided not to try to predict what he'd say, if indeed he'd say anything.

Finally his screen lit up and clicking resumed, slowly this time. *With [a] top-down [approach] we put humankind back in the center [of cosmological theory],* he said. *Interestingly, this is what gives us control.*

In a quantum universe we switch on the light, I added. Stephen smiled, noticeably satisfied to discern a whole new cosmological paradigm on the horizon.

What a wonderful twist, I mused. We started out searching for a deeper explanation of the universe's fitness for life in the physical conditions at the origin of time. But the quantum cosmology we developed to this end suggests we were looking in the wrong direction. Top-down cosmology recognizes that, much like biology's tree of life, physics' tree of laws is the outcome of a Darwinian-like evolution that can only be understood backward in time. The later Hawking propounded that down at the bottom, it isn't a matter of why the world is the way it is—its fundamental nature dictated by a transcendental cause—but of how we got where we are. From this viewpoint, the observation that the universe happens to be just right for life is the starting point for everything else. Interlinking not just gravity and quantum mechanics—the large and the small—but also dynamics and boundary conditions, as well as man's worm's-eye perspective on the cosmos, the top-down triptych offers a remarkable synthesis that weans cosmology off the Archimedean point at last.

We really should be going, Stephen's nurse insisted. Walking across the courtyard to the college gate on Trinity Street, Stephen remembered he'd gotten us tickets for Wagner's *Götterdämmerung* the next evening at the Royal Opera House and asked whether I'd drive us to London—*to mark the end of my battles with God.*

He never went back to his old bottom-up philosophy of cosmology. Something had snapped in him that day when I, back from Afghanistan, walked into his office. Years later, paraphrasing Einstein on the cosmological constant, Stephen told me that looking at his no-boundary genesis from a causal bottom-up perspective was his "biggest mistake." In hindsight, we can see that both Einstein and Stephen were taken by surprise by their own theories. In 1917, Einstein's fixation with the age-old idea of a static universe made him fail to grasp the sweeping cosmological implications of his classical relativity theory. In a similar way, Stephen's deeply entrenched causal thinking about the origin of time blinded him to the new vista his semiclassical no-boundary hypothesis uncovered.

The development of top-down cosmology marked the most fruitful and intense phase of our collaboration. At work or in the pub, in the airport or around late-night campfires, the top-down philosophy became a boundless source of joy and inspiration. In *A Brief History of Time,* the early (bottom-up) Hawking famously wrote, "Even if we do find a theory of everything, it is just a set of rules and equations. What is it that breathes fire into the equations?" The answer of the later (top-down) Hawking was: observership. We create the universe as much as the universe creates us.

FIGURE 50. *Stephen Hawking and the author halfway through their journey, in Stephen's office on the new campus for mathematical sciences in Cambridge. On the bookshelf behind us are the PhD dissertations of Stephen's academic offspring. Underneath those, next to the microwave oven, is the speckled microwave background radiation reaching us from all directions in the sky, forming a sphere all around us—our cosmic horizon.*

CHAPTER 7

TIME WITHOUT TIME

Time present and time past
Are both perhaps present in time future.
And time future contained in time past.
If all time is eternally present
All time is unredeemable.

—T. S. ELIOT, "Burnt Norton"

LAUNCHING THE DARWINIAN REVOLUTION IN COSMOLOGY WAS A
quintessential Hawkingian act. It serves as a great example of the bold,
adventurous, and intuition-driven practice of physics that characterized
much of his later work.

Our earliest writings on top-down cosmology date from the year 2002
and whereas in hindsight we were on the right track, we were really walking on quicksand. Even in later stages, the superposition of spacetimes
that lies at the heart of the top-down philosophy remained difficult to get
a grip on. Did they combine to form an enormous extension of Everett's
universal wave function, a kind of quantum version of the multiverse with
tentacles stretching into all corners of the string landscape? If so, however,
might that grand wave function of the cosmos not be the long-sought
metalaw underpinning all physical theory, relegating observership once
again to little more than a post-selection effect?

Our early top-down ideas were what Jim Hartle once called "ideas for
an idea"—insights that were likely profound and important but needed a
home within a proper physical theory to come to fruition. So we began to
search for firmer ground.

Inspiration came from an unexpected corner. Around that time a second revolution in physics was picking up. This one was brewing on the

desks and blackboards in the offices of string theorists, who while experimenting with hypothetical universes, had discovered that these have weird *holographic* properties.

I first heard about the holographic revolution sweeping theoretical physics in January 1998. As a fresh graduate student I was taking the advanced math course at DAMTP known in the Cambridge jargon as Part III when, at the start of the Lent Term, the faculty put together a special series of research seminars—in light of an important new development, we were told, that, rumors had it, would "change everything."

That sounded exciting, so I decided to slip into the seminar room to listen in on the first lecture. This was still in the old DAMTP building on Silver Street in central Cambridge, in a lecture room with scant lighting, predictably foggy windows, and a large blackboard running across the entire length of the front wall. The room was packed with nearly a hundred theoretical physicists, and the atmosphere was loud and informal. Some people were immersed in heated discussions, others were frantically scribbling equations, and yet others seemed to be just chilling out, sipping their tea.

I was looking for a spot to take it all in when my eyes caught the speaker of the day. I had seen him before—Stephen driving his wheelchair was a familiar sight in Cambridge. But the mere sight of him here in his scientific headquarters revealed a whole new dimension of his persona. Despite his near immobility, Stephen was full of life. Clearly much beloved by his peers and at the epicenter of his gravity group, he was smiling and interacting with those around him in all sorts of subtle ways I couldn't decode. The entire scene exuded an air of kinship and sheer joy. I felt as if I were crashing an extended family party. On the menu: the end of spacetime as we know it.

Stephen was maneuvering his wheelchair, his left hand wrapped around a steering knot on the armrest of his chair, apparently attempting to position himself so that he could see the audience, by moving his eyes up and slightly to the right, as well as the projector screen, by moving his eyes up and slightly to the left. When Stephen was finally happy with the configuration, Gary Gibbons got up and told the audience that Stephen would deliver the first lecture of their special series, and the room fell silent. Stephen, clicker in his right hand, began to execute a series of operations to get a text he had prepared to appear on the screen fixed to his wheelchair. Then he paused, looked up at us, looked back at the screen, and clicked once more.

I have always had a soft spot for anti–de Sitter space and felt it was unjustly neglected. So I'm pleased it has come back in fashion with a bang.

Stephen read his lecture by sending his script sentence by sentence to the computerized voice attached to his chair. In the front row sat an assistant with a printout of the text on his lap. He operated a projector to show a few slides with basic illustrations of anti–de Sitter space and other shapes of space featuring in Stephen's lecture. Sometimes Stephen paused to make eye contact with his audience, gauging our reaction to a joke he was proud of, or to let a controversial statement sink in.

I was mesmerized, first, by Stephen's performance but also by that strange anti–de Sitter space that was the source of so much excitement. Little did I know that barely a year later, Stephen would instruct fellow Hawking student Harvey Reall and myself to think about our visible universe as a four-dimensional membrane-like hologram hovering in a five-dimensional anti–de Sitter space. Together we would write "Brane New World."[1] A layman's version of this tract ended up in *The Universe in a Nutshell*, which we were editing at the time. The way Stephen wove his technical research nearly simultaneously into his books was impressive, almost unheard of in the exact sciences.[2]

As a matter of fact, the idea that the universe may be akin to a hologram has a long history. You might remember Plato's allegory of the cave, in which Plato likens our perceptions of the world to those of prisoners confined to a cave watching shadows meander across the walls. Plato imagined that our world of appearances was only a faint glimmer of a far superior reality of perfect mathematical forms that existed out there, independently of us. Today, the holographic revolution in physics is turning Plato's vision on its head. The latest incarnation of holography envisions that everything in the four dimensions we experience is in fact a manifestation of a hidden reality located on a thin slice of spacetime. Holographic thinking posits there is an alternative description of reality, a completely different way of looking at the world, from which gravity and warped spacetime are somehow projected. Moreover, it holds that this three-dimensional shadow world of quantum particles and fields may be telling the full story after all. In its most ambitious form, twenty-first-century holographic physics asserts that if only we could decode the hidden hologram, we would understand the deepest nature of physical reality.

The theoretical discovery of holography ranks among the most important and far-reaching discoveries in physics of the late twentieth century. It also had an immediate influence on Stephen's thinking, drawing him deeper into string theory. And even though physicists still don't quite agree where exactly the hologram would be located or what it would be made of, the novel vista that holography has revealed has already changed the field of theoretical physics beyond recognition. For decades, theoretical physicists had struggled to complete the unification of general relativity and quantum theory that string theory had initiated. The discovery of holography did exactly that. It showed that gravity and quantum theory need not be water and fire but can be like yin and yang, two very different yet complementary descriptions of one and the same physical reality.

While holography wasn't invented with a realistic universe in mind, cosmology may well be the arena where it will ultimately have its most radical implications. Holography provided the avenue Stephen and I had been looking for to put top-down cosmology on firmer footing. And as I will describe in this chapter, it effectively renders a full-blown top-down approach to unlock the big bang inevitable.

The development of a holographic cosmology marked the third stage of our journey. We got started on this third leg during one of Stephen's visits to Belgium, in the fall of 2011, and it eventually culminated in a paper we published shortly before his passing.[3] Above all, this is a journey deep into the cutting edge of theoretical physics, interlinking far-flung fields, from quantum information to black holes and cosmology, in a tantalizing synthesis that suggests there can be "time without time."

THE FIRST HINTS of holography go back to the golden age of black hole research in the early 1970s, when mathematicians and theoretical physicists finally understood the basic properties of these incredibly dense objects.

This golden age culminated in Hawking's startling discovery that black holes aren't completely black but emit a faint trickle of radiation. At first Stephen famously thought he had made a mistake in his calculations. Black holes were supposed to swallow all matter and radiation, not eject it—after all, that was thought to be the very essence of a black hole. What convinced him that his calculations were correct and the radiation was real was that it had all the characteristics of thermal, so-called *black-body radiation,* physicists' term for the kind of radiation emitted by an ordinary nonreflective body at a given temperature. The 2.7 Kelvin cosmic microwave background radiation, for example, is black-body radiation. It tells us that even the entire observable universe behaves like an ordinary radiating body.

In the year 1900, Planck's theoretical derivation of the spectrum of black-body radiation marked the dawn of the quantum revolution. Today, whenever Planck's spectrum shows up in nature, physicists take this as a telltale sign of an underlying quantum process. This was precisely the sort of process Hawking considered. Stephen looked at black holes from a semiclassical angle, studying the quantum behavior of matter moving about in the classical, warped geometry of a black hole. To his surprise, he found that quantum processes near the horizon surface, the point of no return in relativity, gave rise to a tiny flux of thermal radiation streaming away from the black hole in all directions. And he went on to calculate the temperature, T, of a black hole, producing the formula shown on the medallion in figure 51.

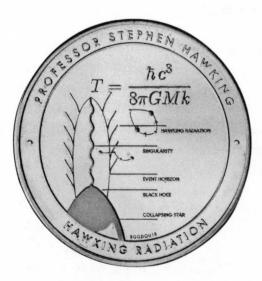

FIGURE 51. *Stephen's formula for the temperature of a black hole, together with a depiction of the process of Hawking radiation, featuring on the medallions struck at the occasion of the interment of his ashes at Westminster Abbey on June 15, 2018.*

The letter M in this formula stands for the mass of the black hole. The remaining quantities are all basic constants of nature: c is the speed of light, G is Newton's gravitational constant, $\hbar$ is Planck's quantum constant, and k is Boltzmann's constant for thermodynamics—the study of energy, heat, and work. The sheer beauty of Hawking's formula is that it brings together all these constants in a single equation. Unlike other celebrated equations of twentieth-century physics, such as the Einstein or the Schrödinger equations, which describe separate domains of physics, Hawking's formula exhibits the interplay of different areas. By combining principles from quantum theory and general relativity, Hawking had taken a mathematical risk, but he was rewarded with an insight that neither relativity nor quantum theory alone could ever have provided: Black holes radiate. Wheeler once said of Hawking's formula that just talking about it was like "rolling candy on the tongue." Today the black-hole temperature equation is inscribed on Stephen's tombstone in Westminster Abbey, as if it were his ticket to immortality.*

Stephen's discovery struck as a bolt from the blue. He announced his result in February 1974, in a stunning talk at a quantum gravity meeting at the Rutherford Appleton labs near Oxford. "Black Holes are White Hot," he declared, leaving his audience dumbfounded. To be sure, this was an archetypical Hawkingian exaggeration. For black holes that are remnants of stars, the numbers in his formula come to a temperature of less than 0.0000001 Kelvin, much colder even than the bitterly cold 2.7 Kelvin CMB radiation. Hence we are unlikely ever to observe black hole radiation. But this is merely a practical inconvenience. Hawking radiation is revolutionary for theoretical reasons alone, since it upends the classical image of black holes as empty, bottomless pits in spacetime out of which nothing can escape.

The reason that it upends this image is that thermal radiation usually originates in the motions of an object's internal constituents. This is why temperature goes hand in hand with entropy, Boltzmann's measure of the number of microscopic arrangements of a system's constituents that leave its macroscopic properties unchanged. Entropy, in turn, is closely related to information, the idea that every matter particle and every force particle in the universe contains an implicit answer to a yes-or-no question.

---

* It isn't the only formula one can find in Westminster. In the nave of Westminster Abbey close by Newton's tomb, there is a memorial stone for Paul Dirac. The inscription on the stone includes the "Dirac equation," $i\gamma{\cdot}\partial\psi = m\psi$, describing the quantum behavior of the electron. Once when I visited the abbey with Stephen, he couldn't resist remarking that "apparently, God was a pure mathematician."

Roughly speaking, higher entropy means that more information can be stored in a system's microscopic details without changing its overall macroscopic properties. Now, from his formula for the temperature of black holes, Hawking could immediately derive an expression for the amount of entropy, S, they contain. Here it is:

$$S = \frac{kc^3 A}{4G\hbar}$$

As a matter of fact, Hawking wasn't the first to propound that black holes have entropy. Already in 1972 the Israeli-American physicist Jacob Bekenstein had advanced the idea that black holes possess an entropy in proportion to the area, A, of their horizon surface. At the time nearly everyone in the scientific community—Stephen up front!—dismissed Bekenstein's idea because, well, black holes don't radiate and hence they can't have entropy. With his discovery of Hawking radiation, Stephen inadvertently proved Bekenstein right.

Bekenstein and Hawking's entropy formula predicts that black holes have a truly gigantic information storage capacity. Black holes are likely the most space-efficient storage devices in the universe indeed. According to their formula, Sagittarius A*, the huge black hole of four million sun masses lurking in the center of the Milky Way—the shadow of which was first imaged in the spring of 2022—can store no less than 10^{80} gigabytes. All data in the Google storage banks, the formula tells us, could easily fit into a black hole smaller than the size of a proton. (Of course, once in, this information would be very hard to google!) Nevertheless, as large as the entropy may be, the formula clearly tells us that the number of bits inside a black hole is finite. The most straightforward reading of the entropy equation is that there is an enormous but finite number of black holes that look the same from the outside but nonetheless differ in their interior constitution.

This is intriguing. According to classical general relativity, black holes are the epitome of simplicity. Relativistic black holes put up the most inscrutable poker face. It doesn't matter whether a black hole is made of stars, or diamond, or even antimatter, Einstein's theory says. In the end, it is fully characterized by just two numbers: its total mass and angular momentum. Wheeler famously summed up this supreme simplicity with the dictum "Black holes have no hair," conveying the idea that black holes appear to retain no memory whatsoever of their own formation history. A black hole in general relativity is the ultimate trash can, with a singularity in the interior that has an infinite capacity to absorb and destroy all information falling into it.

But Bekenstein and Hawking's semiclassical entropy formula paints a very different picture. It portrays black holes as the most complicated objects in nature, the exact opposite of their classical image. The entropy formula suggests that Einstein's general relativity, because it ignores quantum mechanics and the uncertainty principle, completely misses the huge number of gigabytes encoded in a black hole's interior microstructure.

This being said, the fact that the entropy grows like the surface area, A, and not like the volume of a black hole is even more surprising. The information storage capacity of all familiar systems is tied to their interior volume, not to the area of their bounding surface. If one wanted to estimate the amount of information in a library of books, for example, one had better count the number of books on all shelves, not just those lining the walls. Not so for black holes, it seems. To compute the quantum information content of a black hole, the entropy formula instructs us to consider the horizon surface area, A, and to cover this with a gridlike pattern of minuscule cells, the sides of each one measuring one Planck length (see figure 52). A Planck length, l_p, is basically one quantum of length. It is the shortest length scale for which the notion of distance has any meaning. Expressed in terms of the above constants of nature, the area of a single Planck-sized cell reads $l_p^2 = Gh/c^3$, which is about 10^{-66} cm^2. Measuring the surface area of the horizon in quanta of Planck-sized squares, the entropy formula predicts that a black hole's total information content is the number of such cells needed to cover the entire horizon divided by four. So the monumental insight that flows from the entropy equation is that each Planck cell on the horizon carries one bit of information. Each of these bits can potentially provide the answer to a single yes-or-no question about the evolution of the black hole and its microstructure, and the collection of all those bits would be all there is to know about the black hole.

This was the first glimmer of holography in modern physics: The storage capacity of black holes isn't determined by their interior volume but by the area of their horizon surface. It is as if black holes do not have an interior but are *holograms*.

FIGURE 52. *The entropy of a black hole equals the number of Planck-sized cells needed to cover its horizon surface, divided by four. It is as if each such minuscule cell stores a single bit of information, and their totality is all there is to know about a black hole.*

· · ·

WHAT ARE WE to make of all this? The entropy formula doesn't tell us how black holes store their zettabytes, or even whether their quantum chips really are stitched to the unfathomable horizon surface. Neither does the entropy specify the list of yes-or-no questions to which the bits of information that it counts supposedly provide an answer. It merely indicates that these bits should exist.

Things get even more confusing when we imagine what might happen to the hidden information when a black hole grows old. The black hole mass, M, enters in the denominator of the temperature formula. So if a black hole loses mass by slowly radiating energy and particles, its temperature goes up, making the hole shine more brightly and lose mass at a faster rate. Hence Hawking radiation, even though it starts out more slowly than one can even imagine, is a self-reinforcing process that eventually makes black holes disappear. This hadn't escaped Hawking.[4] "Black holes are not eternal," he wrote. "They evaporate away at an increasing rate until they vanish in a gigantic explosion."

But what is the fate of the huge amount of information stored inside when black holes radiate and ultimately evaporate?

There would seem to be two reasonable scenarios. First, the information is lost forever. Black holes are the ultimate erasers. Given the swallowing power of black holes, this may seem like a natural outcome. But the thing is, quantum theory forbids this scenario. The basic rules of quantum theory stipulate that the wave function of any system evolves in an information-preserving manner. Always. Quantum evolution can process information beyond recognition, but it can never irreversibly obliterate information. This property is tied to the obvious requirement that the probabilities in quantum theory must always add up to one, no matter what happens. The preservation of information means, for example, that when you burn an encyclopedia, the laws of quantum physics predict that you can in principle retrieve all information from its ashes. Likewise, if quantum mechanics holds near the horizon surface of black holes—and we have no obvious reason to doubt that it holds—then every scrap of information must ultimately come back out when the black hole eventually disappears.

Let's consider the second scenario. Perhaps all information leaks out, encrypted in the Hawking radiation? Since the evaporation process takes eons, this doesn't seem implausible either. What's more, this would be nicely consistent with quantum mechanics. Alas, this is not what Stephen's calculations say happens. Hawking radiation doesn't carry away any information. When a black hole emits some of its mass in the form of Hawking radiation, the spectrum of that radiation is as featureless as it can be. Nothing about the radiation reveals anything about the hole's microscopics or its past history. Once a black hole radiates its last ounce of mass and disappears, all that's left, according to Hawking, is a cloud of random thermal radiation from which it would be impossible, even in principle, to know whether there ever was a black hole—let alone which one. Evaporating black holes, Hawking proclaimed, are fundamentally different from burning encyclopedias.

This is a paradox. Information appears to be irretrievably lost when black holes evaporate, yet quantum theory says this is impossible. Gradually it dawned on physicists that Stephen, with his ingenious thought experiment, had put his finger on a fantastically profound and difficult problem that arises when relativity and quantum theory venture into the same waters. On the basis of what seemed like a perfectly fine semiclassical amalgam of both theories,[5] he had shown that the abyss separating both theories was in fact far deeper and wider than he or anyone else had anticipated. The paradox of the fate of information locked inside evaporating black holes became the most vexing puzzle in theoretical physics of the late twentieth century, bedeviling not one but two generations of physicists. It is in some ways the contemporary analogue of the Mercury anomaly in the nineteenth century, the wobble in Mercury's orbit that defied Newton's theory. As such the black-hole information paradox became a beacon in the search for a unified theory. Physicists felt that if they could disentangle Hawking's knot and understand what happens to the hidden information when black holes cease to be, they'd be well on their way to marrying the principles of relativity and quantum theory into a single coherent framework.

EARLY ON, STEPHEN bet on the first scenario: Information is lost; physics is in serious trouble; quantum theory must be revised. "Breakdown of Predictability in Gravitational Collapse" was the title of the paper in which he first elaborated on the consequences of information loss.

To be sure, a black hole with the mass of the sun won't start evaporating until a few hundred billion years from now, when the temperature of the microwave background radiation finally falls below that of stellar black holes. The evaporation process itself will then take another 10^{60} years at least, much longer than the current age of the universe. So, unless the hot big bang already produced mini black holes or unless the Large Hadron Collider at CERN were one day up to fabricate these after all, black hole explosions will likely remain theoretical thought experiments for quite some time.

But Stephen's was a point of principle. If black holes destroy information, then they could emit any collection of particles whatsoever when they finally begin to evaporate. This would mean that the life cycle of black holes, from the gravitational collapse of a star to a cloud of Hawking radiation, would imbue physics with a whole new level of randomness and unpredictability, on top of the usual probabilities of quantum mechanics. It would be as if part of the wave function of a collapsing star simply vanished inside black holes, or somehow leaked off, perhaps, into another universe. Obviously this would jeopardize the ability of physics to predict the future of *our universe,* even in the reduced probabilistic sense familiar from quantum mechanics. And if determinism, the probabilistic predictability of the universe on the basis of scientific laws, were to break down in the presence of black holes, how could we be sure it doesn't break down in other situations? How could we be sure of our own history, and our memories? "The past tells us who we are," Stephen noted pointedly.[6] "Without it, we lose our identity." Contemplating the far-reaching consequences of the loss of information inside black holes, Stephen was led to conclude that physics was in serious trouble indeed.

For years the argument went back and forth without much progress. Those who came at the problem from the angle of particle physics argued that quantum theory stood firm and that Stephen had made a mistake. Yet no particle physicist could find an error in Stephen's calculations. Most relativists, on the other hand, acutely aware of the sheer destructive power of spacetime singularities, sided with Stephen but failed to come up with a convincing strategy to rescue physics. The upshot was a stimulating scientific environment that entangled both research communities. Particle physicists and relativists, employing different methods and tools, began to learn from one another, united in their search for a deeper truth hidden in the faint photons that make black holes shine.

But it wasn't until the dawn of the twenty-first century, when physicists finally got a better grip on the holographic nature of black holes, that a whole new array of thoughts and thought experiments broke the impasse on the black-hole paradox. These insights arose out of the so-called second revolution of string theory, the theory that in the late 1990s had given a boost to multiverse cosmology and played a central role in physicists' efforts to formulate a unified quantum theory of gravity and all other forces (see chapter 5).

THE PREEMINENT STRING theorist Edward Witten of the Institute for Advanced Study in Princeton fired the first shot of the second string revolution, with his lecture at Strings '95, the 1995 edition of the annual gathering of string theorists worldwide.

It must be said that string theory wasn't in great shape at the time. The prospects appeared poor (to put it mildly) that physicists would ever be able to test any of the core ideas of the theory. The highest-energy particle collisions at the world's largest accelerators had shown no sign—and they still haven't to this very day—that there exist curled-up extra dimensions into which some of the energy released in collisions leaks. The ultra-tiny Planck scale, where the quantum nature of gravity would surely become important, seemed completely out of reach, for you would need a particle accelerator as large as the solar system to probe scales that small. Furthermore, despite years of innovative mathematical wizardry, the theory had failed to shed light on the quantum nature of gravity in situations where it really mattered—inside black holes and at the big bang. And to make matters worse, string theorists had realized that there wasn't just one string theory but five different variants, which all laid an equal claim to be "the" unified theory of nature. On top of this there was a heretical sixth theory, called supergravity, an extension of Einstein's relativity that involved matter and supersymmetry and contained membrane-like objects, not strings. In fact, as a hotspot of supergravity, Cambridge had developed a bit of an anti-string reputation in this period.

"Some Comments on String Dynamics," the title of Witten's lecture at Strings '95, didn't suggest he was about to break this stalemate. But he did exactly that. In what would go down in the annals of physics as a legendary lecture, Witten sketched a radical new perspective on string theory. He expounded that the five string theories and the dissident supergravity theory weren't six separate theories but merely different faces of a single mathematical edifice. Combining a wide array of insights, Witten identified a sophisticated network of mathematical relationships that morph the various string theories into one another and into supergravity, creating a weblike entity interlinking them all (see figure 53). He dubbed this web *M-theory*. And even though M-theory may not have a definite structure in and of itself—some claim M stands for magic or mystery—it has this amazing shape-shifting ability, somewhat like a boggart really, by which it acquires the form of one of the six partner theories, depending on one's perspective. This deeper union that M-theory uncovered was enough to spark the second string revolution. M-theory made theorists realize that their six different approaches toward a unified theory weren't in conflict with each other but complementary inroads into the realm of quantum gravity that could reinforce one another.[7]

Mathematical relationships that transform seemingly distinct theories

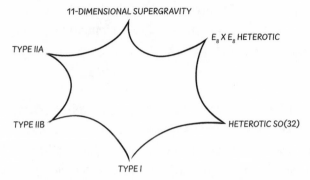

FIGURE 53. *A web of mathematical relationships interlink the five string theories and supergravity theory, hinting at a deeper unifying story.*

into each other are known to physicists as *dualities*. Two theories that are dual are equivalent in some fashion; they describe one and the same physical situation expressed in a different mathematical language. A simple example is the wave/particle duality in quantum mechanics, which caused a great deal of confusion in the early days of the theory.

Dualities are powerful calculational assets, for by offering a complementary perspective on a given physical system, they can unlock new insights. The dualities of M-theory are especially powerful because they often transform a daunting analysis in one string theory into a straightforward problem in its dual partner theory. Before the second string revolution, physicists had to rely on approximate methods to analyze each of the string theories separately. This restricted their playing field to semiclassical situations in which a relatively small number of strings were vibrating in a gently curved classical background space. As a result, the fascinating quantum properties of black holes, let alone of the big bang, remained out of reach of their analyses, and the grand project of unification remained bogged down. The second string revolution dramatically changed all this. Ever since, when the going gets tough in one string theory, a duality has often come to the rescue to reshuffle an impossibly difficult calculation into a perfectly feasible one in a different string theory. So, Witten's M-theory web is much more than the sum of its component theories. By stitching insights from all five string theories and from supergravity, M-theory has opened up wholly uncharted territories in the quantum realm of gravity and unification.

But the pinnacle of the second string revolution was the discovery of a duality of an entirely new kind, a duality so strange that no one thought it could exist—a *holographic duality*.

IN 1997 ARGENTINIAN-BORN Juan Maldacena, while working as a young assistant professor at Harvard University, hit on a most intriguing duality that linked neither two string theories, nor two particle theories. Instead, it related string theories with gravity to particle theories without gravity. What's more, the two sides in Maldacena's duality live in a different number of dimensions: The particle theory is like a hologram of the gravity theory.

Maldacena uncovered this strange duality by thinking about string theory and supergravity in a specific hypothetical setting.[8] The gravity side of Maldacena's duality involves general relativity and supergravity in universes with a shape similar to anti–de Sitter space, or AdS space. As its name suggests, AdS space is the antipode to de Sitter space. The latter is the solution to the Einstein equation found by the Dutch astronomer Willem de Sitter in 1917 that describes an exponentially expanding universe filled with a positive cosmological constant $[\lambda > 0]$. Anti–de Sitter space has a negative cosmological constant $[\lambda < 0]$ and does not expand. Quite the contrary, it is somewhat akin to the interior of a snow globe world, a spherical box bounded on all sides by an impenetrable surface.

The second side of Maldacena's duality involves quantum theories of particles much like the Standard Model. These are quantum field theories, or QFTs, because they describe particles and forces as localized excitations of spread-out fields. The QFTs in Maldacena's duality are similar to quantum chromodynamics, the part of the Standard Model that describes the strong nuclear force.

The surprising holographic nature of this duality comes about because the quantum fields on the particle side don't penetrate into the interior of the AdS snow globe world but can be thought of as being confined to the boundary surface that surrounds it. So the QFT apparently operates in a spacetime with one dimension less. If the AdS space has four spacetime dimensions, then the QFT lives in three dimensions. It lacks the interior depth of AdS, the curved dimension that runs perpendicular to its boundary surface. The QFT is also gravity-free. On the boundary of AdS space there are no gravitational waves, black holes, or even anything resembling a gravitational attraction. Gravity does not exist in a QFT of particles.

Or so we thought. The crux of Maldacena's audacious claim was that these two theories, however different they might seem, were, in fact, disguised versions of each other. Maldacena argued that (super)gravity theory in AdS and QFT on the boundary were in some sense equivalent. This is holography in action! For it would mean that everything there is to know about strings and gravity in a four-dimensional AdS universe can be encrypted in quantum interactions of ordinary particles and fields lying entirely in the three-dimensional boundary surface. The surface world would function as a kind of hologram, as a blueprint of the interior AdS world that contains all the information but differs profoundly in appearances. It is almost as if you could learn everything about the interior of an orange by meticulously analyzing its skin.

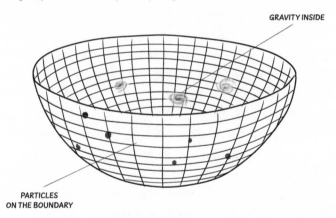

FIGURE 54. *Holographic relationships equate string theory and gravity in the interior of a curved spacetime to certain quantum theories of particles and fields without gravity living on the boundary of that spacetime.*

In its most ambitious form, the holographic duality states that a boundary world of quantum fields and particles fully specifies the behavior of gravity and matter inside AdS, and not just a classical or semiclassical approximation of it. What makes this doubly exciting is that the particle theories popping up in Maldacena's duality are among the best understood quantum field theories, having been studied in depth by particle physicists since the middle of the twentieth century. Hence holography—

in its ambitious form—provides a *working example* of a complete quantum theory of gravity and matter.

This was a real game changer. For decades physicists had struggled to seal the marriage of general relativity and quantum theory. Ever since Maldacena's epiphany, these two seemingly conflicting theories have been working in symbiosis. Holographic dualities have revealed that relativity and quantum theory aren't antagonists but merely alternative vantage points on the same physical reality. Physical systems can be gravitational and quantum at the same time, holography says, albeit in different dimensions. Such was the stunning change of perspective brought about by Maldacena's duality.

In line with other dualities in M-theory, the nature of the relationship between both sides of a holographic duality is such that when calculations on one side are perfectly straightforward, the situation on the other side will often be thoroughly complicated. When gravity is weak and the AdS universe is gently curved, for example, the boundary description involves such strong quantum interactions between its constituents that the QFT becomes utterly intractable and even the notion of individual particles may cease to have much meaning.

This property makes holographic dualities very hard to prove but also extraordinarily powerful. For it means that physicists can use Einstein's gravity theory, and its extension to supergravity, to learn about new phenomena in the quantum world of particles, and vice versa. Over the years holography has become a veritable mathematical laboratory, in which theorists have performed the most ingenious thought experiments in order to gain a better understanding—and some intuition—of nature's fascinating holographic underpinnings. Today, holographic physics has spread its wings well beyond its M-theory origins, with a rich network of relations interconnecting what we used to think of as disparate branches of physics, from general relativity, condensed matter physics, and nuclear physics, to quantum information, and even astrophysics.

BUT LET US return to black holes. If holography amounts to a full theory of quantum gravity, albeit in an AdS setting, then surely it resolves Stephen's notoriously thorny black hole information paradox?

Well, it's rather subtle. The reason is that Maldacena's surface description encrypts the interior AdS world in a highly scrambled and utterly unrecognizable manner. This shouldn't come as a surprise; even an ordinary optical hologram doesn't remotely resemble the three-dimensional scene that it encodes. The surface of a common two-dimensional hologram contains lines and scribbles that seem random. It requires a complex operation, usually in the form of a laser light shining through the surface, to convert these into a three-dimensional scene.

Likewise, it requires a sophisticated mathematical operation to decipher what's happening inside AdS space from the holographic surface description. Unfortunately, the discovery of holography didn't come with a mathematical dictionary that we can turn to to find out how the two sides translate into each other. Theorists have had to develop this dictionary, entry by entry, in order to decode the hologram and thereby unlock the sheer power of the holographic dualities.

Perhaps the very first entry that you'd be looking for in the AdS–QFT dictionary concerns what is arguably the duality's strangest property: the disappearing dimension. How do particles and fields confined to a surface nevertheless capture all that's happening in the interior depth of AdS? Every scrap of information about everything inside the AdS universe must somehow be encoded in the QFT, because otherwise the duality wouldn't be a duality. So how do quantum field theories manage to somehow "absorb" an entire dimension?

The key property of AdS that is relevant here is that the interior dimension perpendicular to the boundary surface is highly curved. The "anti" in anti–de Sitter space refers to the fact that AdS space has negative curvature, meaning that the angles of a triangle add up to less than 180 degrees. (On the positively curved surface of Earth, or in de Sitter space, the angles of a triangle add up to slightly more than 180 degrees.) The negative curvature means that a projection of AdS on a flat surface produces an anti-Mercator effect: Regions near the boundary look too small (as opposed to too large on a Mercator map of Earth's surface). A two-dimensional spatial cut running across the AdS interior projected onto a flat surface looks much like *Circle Limit IV*, M. C. Escher's famous woodcut of a disk with endlessly repeated figures of angels and demons (see figure 55). In the true negatively curved AdS space, all angels and demons have the same size. But in Escher's flattened projection the figures get smaller and smaller and pile up near the circular boundary, fading away in an infinite fractal at the edge.

FIGURE 55. *M. C. Escher's* Circle Limit IV.

Now, if you imagine projecting one of the angels (or demons) in Escher's woodcut onto the circular boundary of the disk, say by creating a shadow in the form of a line interval, that line will be much shorter for an angel located near the edge than for the same angel residing deep in the interior. And this is exactly how holography works: Maldacena's duality translates "interior depth" in AdS into "size" on the boundary. Thus, the very first entry in the AdS–QFT dictionary reads that shrinking and growing in the boundary world corresponds to moving in the direction perpendicular to the boundary in the curved AdS universe, respectively approaching or receding from the edge.

In fact, the idea that in quantum field theories, scaling things up or down is like moving about in an additional dimension has a long history. Size is closely related to energy in particle physics. The reason that particle physicists ask for larger accelerators is that by raising the energy of particle collisions one can probe nature at smaller distances. It is like buying a better microscope. Now, crucially, the collection of particle excitations and force interactions that a given QFT describes depends on the distance resolution one has in mind. The particle content of a QFT employed at low energies, or large length scales, can be very different from the particles and forces that come to the fore in the same theory at high energies. So the basic quality of size, or equivalently, energy, in quantum field theories stores additional information. In the mid-twentieth century, physicists developed the mathematical formalism that prescribes exactly how the properties of a given quantum field theory change as one varies the energy scale at which it is used. Maldacena's duality cleverly exploits this property. The AdS–QFT dictionary translates this abstract "energy dimension" of QFTs into a "curved dimension of space" on the gravity side.

But what about the no doubt fascinating "black hole" entry in the AdS–QFT dictionary?

Within a few months of Maldacena's paper, Witten had put a black hole in the AdS interior, hopped over to the boundary theory, and looked at its hologram. Since there is no gravity in the boundary world—at least not in a familiar sense—we shouldn't expect the hologram of a black hole to bear any resemblance to the bottomless spacetime pit of Einstein's relativity. And indeed it doesn't. When Witten investigated the dual description of a black hole, he found little more than a swarm of hot particles. Holography apparently transforms the most enigmatic objects in the universe into something rather ordinary. The holographic story of the life cycle of black holes, the cycle that has proven so difficult to grasp in the language of gravity, reads something like the heating and subsequent cooling of a plasma of hot quarks and gluons, a process that is hardly more exotic than what experimental physicists create on a daily basis in their laboratories by slamming heavy nuclei into each other. Furthermore, the thermal entropy of a hot quark stew on the boundary surface equals the entropy of a black hole in the AdS interior, obviously an important test of the holographic duality. In fact, the mathematical observation that the black hole entropy grows like the surface area of the horizon is no longer surprising in the light of holography, for the horizon surface and the quark stew live in the same number of dimensions.

Almost as an afterthought, as if it were a footnote to the black hole entry, Witten remarked that the surface description of the formation and evaporation of black holes appears consistent with quantum theory. The holographic duality did appear to resolve Hawking's paradox. The reason is that the fairly ordinary clusters of particles that make up the dual description of black holes have wave functions that evolve in a smooth and information-preserving manner, according to the usual quantum rules *without* gravity. While the quantum dynamics of hot quarks may scramble and transform information, we know for certain that it doesn't destroy information since this isn't even an option in a QFT. By the logic of the duality, then, all information inside evaporating black holes in an AdS universe must eventually leak out and wind up in the emitted Hawking radiation.

NOW YOU MIGHT have thought that Maldacena's and Witten's discoveries made Stephen swiftly change his spots on the fate of information inside black holes. They did not.

Why not? Because Witten's argument didn't quite complete the information paradox entry in the AdS–QFT dictionary. Witten's duality-based reasoning that all basic bits inside a collapsing star ultimately survive is highly formal. It doesn't explain *how* information winds up in the Hawking radiation. The duality only says that, somehow, it does. If toward the end of 1998 a fearless astronaut had called Princeton to double-check whether he could get out of a black hole, the local theorists would have said "Yes, sure, you will just be very scrambled." But if he had pressed them and asked how to do it, Witten and his colleagues would have had to admit they had no clue. The gravity description of the escape from an old, evaporating black hole remained deeply mysterious in the early years of holographic physics. Maldacena's awesome duality successfully removed any formal contradiction between quantum theory and black holes, but it hardly elucidated where Stephen had made a mistake in his original gravity-based calculations. Understandably, then, and to his credit, Stephen insisted on a resolution of the paradox on his own terms: a description of the escape route in the language of gravity and geometry that didn't require him to trust the dual magic blindly.

It would be another six years before Stephen finally came around and publicly declared quantum mechanics safe in the presence of black holes. He did so with ample drama. The chosen venue was the 17th International Conference on General Relativity and Gravitation held in Dublin in July 2004, the same sort of gathering at which in 1965 he had first presented his big bang singularity theorem. When Stephen emailed the conference conveners to request a speaker's slot because "he had solved the black hole information paradox," they not only gave him one but booked him in the main concert hall at the Royal Dublin Society. Soon afterward they were dealing with a shortage of press passes for what was supposed to be a scientific lecture.

As usual the conference was an occasion for the Hawking clan of students and former students to catch up. On the eve before Stephen's lecture we went out for a drink in Dublin's Temple Bar. Savoring a rare leisurely moment, Stephen turned up the volume on his speech synthesizer. *I'm coming out,* he declared with a big smile. And indeed, the next day Hawking told a hall packed full with an unusual mix of physicists and journalists that black holes weren't the bottomless pits he once thought they were but instead release all there is to know about their past when they radiate and disappear. During the press conference that followed his lecture, Stephen paid off a bet with the eloquent Caltech physicist John Preskill, who in 1997 had wagered Stephen and Kip Thorne that all information eventually leaks out from evaporating black holes. The bet stipulated that "the loser(s) will reward the winner(s) with an encyclopedia of the winner's choice, from which information can be recovered at will." Stephen presented John with a copy of *Total Baseball: The Ultimate Baseball Encyclopedia,* though not without mentioning that he should perhaps have given him the ashes. John euphorically raised the encyclopedia above his head as if he had just won the World Series. Flashbulbs popped, and one of those pictures wound up in *Time* magazine.

Yet Stephen's Dublin performance was a somewhat awkward act. Of course we had long been used to the fact that his every thought on black holes acquired a life of its own in the public arena. Brilliant at communicating to a worldwide audience—and having been in tune with popular culture since childhood—Stephen had become one of the great voices of science of our age, inspiring millions all over the world. But the Dublin act was a rare occasion when the line got blurred between Stephen's public image and proper scientific practice. For despite the media hype surrounding Stephen's U-turn on black holes, neither his Dublin lecture nor his subsequent paper on the subject much advanced the matter—let alone settled it. Most string theorists in attendance had concluded six years before that black holes don't destroy information, and they felt Stephen's concession speech was long overdue. Relativists, on the other hand, weren't swayed by Stephen's abstruse presentation and felt he had changed his mind prematurely. Among them was Kip Thorne, who refused to concede their bet in Dublin—and I don't think he ever did.

Stephen had been pursuing his fresh attempt at straightening out the black hole information paradox together with the gallant Frenchman Christophe Galfard, his student at the time, who had had the (mis)fortune to walk through the door into Stephen's office in a "black hole year." Christophe too realized that their calculations weren't working out quite as neatly as they had anticipated but rather pointed to deeper-lying questions still. So why did Stephen take to the stage in Dublin to declare that information isn't lost inside black holes? What made him feel that despite the absence of a firm proof, the body of evidence had nonetheless shifted toward information preservation? I think he had his eye on an obscure and underappreciated element of holography that he felt was the key to sorting out the paradox, i.e., that there's more than one interior.

You see, the surface hologram encrypts not just one interior curved geometry but a mélange of different shapes of spacetime.[9] The holographic duality, that is, apparently builds in that radical quantum thinking about gravity, à la Feynman, which I discussed in the previous chapter and which proved crucial to untangling the cosmological information paradox. Holography reinforces these ideas and predicts that at some level, gravity involves not one but a superposition of spacetime geometries. It encourages us to think of the AdS interior as a wave function, not as one single spacetime.

"The moment we say that a black hole is described by Schwarzschild's geometry we have an information loss problem," Stephen told his audience in Dublin.[10] And he then continued: "However, information about the exact state is preserved in a different geometry. The confusion and paradox arose because we thought classically, in terms of a single objective spacetime. But Feynman's sum over geometries allows it to be both geometries at once."

This was the new top-down Hawking talking.

In his original derivation of Hawking radiation, the bottom-up Hawking had (very reasonably) assumed that any radiation that escaped was moving about in the spacetime of the black hole, the warped geometry found by Schwarzschild in 1916. Of course, such an assumption precludes the possibility that in the long run, a whole new shape of space would come into play. Thirty years on, Stephen saw that this reasoning had been a tad too classical. He now proclaimed that when black holes grow old, surprisingly, much of the information about the hole and its history is no longer stored in the original black hole geometry but in a different spacetime altogether. So the new top-down Stephen conceded— perhaps grudgingly, who knows—that his younger alter-ego had erred before he even began calculating, when he had assumed spacetime to be a given.

In hindsight, top-down Stephen was right in his intuition that there was another geometry involved. Proper quantum thinking in terms of a sum over interior geometries instead of a single one ultimately proved to be a key to begin to untangle the black hole paradox. The controversy over his Dublin lecture lay in the fact that Stephen didn't identify in what curved shape the past of an old black hole might then be stored. He effectively suggested (incorrectly) that the possibility that there wasn't a black hole to begin with was enough to resolve the paradox.

IT WOULD TAKE much more work in Maldacena's holographic laboratory, and the exploration of many more blind alleys, before theorists finally began to discern the escape route out of an old black hole. As a matter of fact, in the years since Stephen's passing, a new generation of black hole physicists, steeped in holography, have realized that perhaps wormholes are involved. Wormholes are exotic shapes of space, somewhat like handles, that function as geometric bridges connecting otherwise far-flung places or moments in spacetime. Figure 56 shows Wheeler's very first drawing of a wormhole, in 1955, which at the time he called a "multiply connected space." Now, in 2019, Geoff Penington, working solo at Stanford, and the Princeton–Santa Barbara string quartet comprising Ahmed Almheiri, Netta Engelhardt, Donald Marolf, and Henry Maxfield found striking evidence that halfway through their evaporation process, black holes might undergo a baffling rearrangement.[11] Their calculations indicated that the slow but gradual accumulation of radiated particles can eventually activate a latent wormhole geometry in the Feynman superposition, creating some sort of geometric tunnel through the would-be horizon region that provides a passageway for information inside to escape.[12]

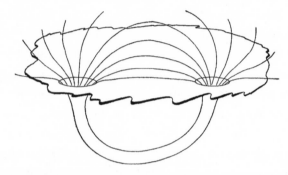

FIGURE 56. *The first schematic representation of a wormhole, drawn by John Wheeler, who coined the term in 1957 to describe tunnels in the geometry of spacetime connecting two distant points. In recent years theorists have suggested that wormholes may provide a passageway for information to escape from an old, evaporating black hole.*

The distant radiation is thought to achieve this remarkable feat by exploiting a subtle quantum phenomenon called *quantum entanglement.* Remember that Hawking radiation originates in quantum jitters of fields near the horizon of black holes. These jitters give rise to particle-antiparticle pairs. Whenever the antiparticle falls into the black hole, its partner particle can escape into the distant universe, where it shows up as Hawking radiation emitted by the hole. However, despite their distance the pairs of particles and antiparticles maintain a quantum mechanical link with each other. Physicists say they remain "entangled." The entanglement means that if you measure the emitted radiation on its own, it looks like random thermal radiation. But if one were able to consider the members of the pairs jointly, one would find that they do contain information, encoded in fine correlations that interlink their individual properties. It's like encrypting your data with a password. The data without the password is nonsense. The password (if you have picked a good one) has no meaning either. But together they unlock information. What Pennington and the string quartet have found—and many theorists have elaborated on since—is that the buildup over eons and eons of more and more quantum entanglement between the inside and the outside of an evaporating black hole can be thought of as generating a wormhole stretching across the horizon. It is as if the particles of Hawking radiation together with their antiparticle partners behind the horizon collectively sew their own space-time bridge, transforming an old black hole from a hermit kingdom into a sort of drive-through.

What's more, quantum entanglement appears to be a key part of how Maldacena's holograms work in general. This touches on what is perhaps the most obvious and at the same time the most profound entry in the AdS–QFT dictionary: Gravity and curved spacetime are emergent phenomena. Years of research has shown that in order for a surface hologram to encode a curved interior geometry, it isn't nearly enough to have a boundary surface with an immense number of particle-like constituents. Instead, a curved interior emerges only if quantum entanglement interconnects numerous boundary constituents. Strikingly, quantum entanglement appears to be the central engine that generates gravity and curved spacetime in holographic physics. It is for Maldacena what laser light is for an ordinary optical hologram.

This is a striking insight. Einstein showed that gravity is a manifestation of warped spacetime. Holography goes further and postulates that warped spacetime is woven from quantum entanglement. Much as the second law of thermodynamics arises from the statistical behavior of numerous classical particles, or sound waves arise from the synchronized oscillations of molecules of matter, the holographic duality radiates the view that Einstein's general relativity emerges from the collective entanglement of a myriad of quantum particles moving about in a lower-dimensional boundary surface. Neighboring regions in the AdS interior correspond to highly entangled components on the boundary surface. Distant parts of the interior space correspond to less entangled portions of the boundary. If the surface configuration has an orderly pattern of entanglement, a nearly empty interior emerges. If the surface system is in a chaotic state, with all its components entangled with one another, the interior contains a black hole. And if one performs an extraordinarily complex quantum operation on the entangled qubits in the hope of reading the black hole's history, one will, bafflingly, find an interior wormhole geometry.

There is a conspicuous element of top-down in all of this. In the language of the previous chapter, we might say that the entangled boundary bits play the role of observership. In top-down cosmology, data on a surface of observation select a past out of an ocean of possible pasts. Holography predicts that in a similar manner, patterns of entanglement on a spherical boundary surface determine the shape of an interior dimension. Both holography and top-down cosmology thus promulgate a striking reversal of the usual order of things in physics: Warped spacetime comes second to the "questions asked" on some boundary surface.

These days there are "quantum gravity in the lab" conferences where gravity theorists and quantum experimentalists discuss ways to create strongly entangled quantum systems made up of trapped atoms or ions, which mimic some of the properties of black holes. By experimenting with these systems, one hopes to learn more about what sort of entanglement patterns underpin warped spacetime and what happens to the geometry when the quantum entanglement that holds it together breaks up. These are very exciting developments indeed. Who would have thought, at the dawn of the holographic revolution in the mid-1990s, that quantum experimentalists would deliver keynote lectures on black hole toy models at the Strings conferences of the 2020s?

It is unfortunate that Stephen didn't live to enjoy these startling new insights. He would surely have been thrilled to see wormholes emerge as the elusive escape channels from evaporating black holes. We can only wonder what pithy one-liner he would have come up with. He would have been equally excited, I believe, to glimpse yet another level of connection between our understanding of black holes and that of the early universe, the two topics that always drove his work. Throughout his research career, insights from black holes often informed his subsequent work on cosmology, from Penrose's black hole singularity theorem to his own discovery of Hawking radiation. The advent of holography led to an even closer interplay between both strands, with cosmological insights such as the top-down approach we began to develop in 2002 inspiring his work on black holes in 2004.

This being said, some string theorists are bemused by these recent advances in the quantum theory of black holes. Their hope has always been that the resolution of the black hole information paradox would replace Hawking's quirky semiclassical mix of geometries with something entirely different. Instead, it now appears that we should take seriously Hawking's superposition of geometries and that, when taken seriously enough, this way of thinking about quantum gravity exceeds everyone's expectations (except Hawking's, of course, which had always been sky-high). Even though much remains to be learned before we will be able to recount a black hole's history by reading its ashes—the Hawking radiation—many theorists now agree that there is no longer a real paradox. Moreover, I'd argue that this development *is* something entirely different. Moving from a single spacetime to emergent spacetimes has truly foundational implications.

For a start, this shift signals the end of the old reductionist dream in fundamental physics. Reductionism is the extraordinarily successful idea that explanatory arrows in science always point downward, toward lower levels of complexity. It holds that in the many-storied tower of science, from physics to chemistry to biology, phenomena at higher levels can in principle be explained in terms of phenomena at lower levels. Reductionism does not mean that lower-level explanations are always needed or useful, or even achievable in practice. Nor is it in conflict with the emergence of new phenomena and "laws" at higher complexity levels. All reductionism means is that such higher-level laws aren't disjunct from their lower-level roots; we qualitatively understand biological phenomena in chemical terms, and chemical phenomena in physical terms. And if we had powerful enough computers to simulate complex biological systems at the microscopic level of their molecular chemistry, we'd expect to see their biological behavior emerge indeed.

But what about the absolute lowest level of the fundamental laws of physics? Is this the rock-solid foundation—the structure of absolutes—that underpins all higher levels of the tower of science? Holography paints a quite different picture. If entanglement, that spooky phenomenon that infamously troubled Einstein—and that was the subject of the 2022 Nobel Prize in Physics—is central to building spacetime, then reductionism versus emergence would appear too limited a way of looking at the world. Holography ingrains a fundamental element of emergence into the very roots of physics—into the fabric of spacetime itself. The holographic duality embodies the view that physical reality and the "fundamental" laws it obeys emerge from a confluence of basic building blocks *and* the way they are entangled. It creates a sort of closed loop of interdependencies that circles from reduction to emergence and back. Holography holds that even the most elementary law-like regularities are ultimately grounded in the complexity of the universe around us. Which leads us to ask: What are the cosmological implications of this conclusion?

AFTER MALDACENA'S DISCOVERY of the holographic nature of anti–de Sitter space, theorists were quick to speculate that our expanding universe too might be a hologram. In the notebooks in which I transcribed some of my conversations with Stephen, I find musings about a possible surface description of expanding de Sitter space as far back as February 1999. But it wasn't until we had our top-down approach firmly on track, more than ten years later, that we began to pursue the idea of a holographic cosmology in earnest.

Unfortunately, by then Stephen was losing the slight control over his muscles that, miraculously, he had managed to retain for so many years of ALS. For reasons yet poorly understood, the long nerve cells that transfer electrochemical signals from the brain to the spine and from the spine to the muscles wither and die in ALS patients, causing their muscles, deprived of any orders, to atrophy. By now ALS had claimed almost all of Stephen's muscular control. Obviously this severely reduced his operational freedom. In the early stages of our collaboration, Stephen could easily steer his wheelchair around to seek out his peers and, clicker carefully placed in his right hand, engage in conversation. At this point Stephen could no longer cruise around independently, which in practice meant his scientific interactions had become limited to a much smaller circle of close colleagues. Moreover, the progression of the disease had made it impossibly difficult for Stephen to operate Equalizer with his clicker. The old-fashioned device, for so many years the umbilical cord connecting his mind to the outside world, from talking and emailing to phoning or googling, was replaced with a sensor mounted on his glasses that he could activate by slightly twitching his cheek. This ensured a vital communication line, but it didn't restore his ability to drive or even to have discussions over lunch or dinner, for example, which previously had been key forums of exchange with his wider circle of colleagues. (In the clicker era Stephen liked to quip that he could eat and talk at the same time.) So, Stephen was at constant risk of becoming isolated. Arguably it was his inability to communicate fluently in his later years that proved the biggest constraint in his scientific life. It meant he couldn't participate fully anymore in those heated debates, over everything from a minus sign in an equation to the merits of philosophy, that we all need to refine and test our ideas. Though all of his cognitive abilities remained intact, during the last decade or so of his life, he became, at times, almost completely locked in.

To make matters worse, he had developed difficulties breathing, and we all feared that he would soon no longer be able to move around at all. But then his support team fitted a ventilator on his wheelchair, and that became a sort of blend of a mobile IC unit and an IT hotspot. Soon, he was on the road again. Moreover, his affluent friends put their jets at his disposal to fly him around the globe, which made for smooth traveling compared to our past expeditions. He would often go to Houston, because he had developed a friendship with the Texas oilman George P. Mitchell, who had taken the initiative to invite Stephen and his circle of close colleagues for an annual physics retreat on his ranch "to create an environment where Stephen could work." And that's what Stephen did. Far away from the bustle of his Cambridge office, year after year I witnessed Hawking's inquiring spirit come alive in the Texas woodlands. It was at Mitchell's ranch, where blackboard sessions seamlessly transitioned into dinner and campfire discussions, that Stephen's holographic theory of the universe was born.

THE VERY FIRST obstacle to apply holography to cosmology is that we don't live in an anti–de Sitter snow globe world. We live in an expanding universe more like de Sitter space. Viewed classically, AdS and its de Sitter antipode have very different properties. The negative curvature of AdS space creates a gravitational field that pulls objects together, toward the center of the space. By contrast, the positive curvature of an expanding de Sitter universe causes everything to repel everything else. This difference can be traced to the sign of λ, the cosmological constant, aka the dark energy term in the Einstein equation. A universe like ours has positive λ, causing it to stretch, whereas AdS space has negative λ, leading to an additional attractive pull. What's more, unlike AdS, expanding universes may not even have a boundary surface that could play host to a hologram. Some expanding universes are hyperspherical, three-dimensional versions of a sphere. Hyperspheres don't have a boundary on which we can hope to encode what's happening inside. Hence it appears virtually impossible to engineer something like Maldacena's holographic duality.

But what if we abandon classical thinking and adopt a semiclassical viewpoint instead? What if we conceive of AdS and its antipode in imaginary time? After all, the principal motivation to develop a holographic cosmology is to get a better handle on the quantum behavior of the universe, and Stephen had long held that geometries with four space dimensions encapsulated its quantum properties. This was the crux of his Euclidean approach to quantum gravity (see chapter 3). Remember the circle he asked me to draw in the hospital (see figure 25)? That circle represented the edge of the disk that you get when you project the quantum evolution of the circular, inflating universe shown in figure 23(b) onto a flat plane. Figure 57 evokes this projection in a more elaborate manner. The no-boundary origin of the universe lies at the center of the disk, where time has morphed into space. The universe today corresponds to the circular boundary. If I were able to draw all four large dimensions, the one-dimensional boundary circle in figure 57 would be a hypersphere, namely the three-dimensional surface in four-dimensional spacetime to which all our observations of the universe are roughly confined. Now, we can see that in this flat projection, the expansion means that most of the spacetime volume that makes up our past gets squeezed toward the edge of the disk. As a consequence, the vast majority of stars and galaxies are piled up near the boundary surface. Does this ring a bell? It does! Replacing stars and galaxies with angels and demons, the disk in figure 57 transforms seamlessly into the Escher-like projection of AdS space depicted in figure 55.

This was exactly the connection Stephen was after. Classical AdS space is not at all like an expanding universe. But from a semiclassical perspective, moving into imaginary time, we see that the two shapes of space are in fact closely related. In the semiclassical realm both AdS and its de Sitter antipode can be thought of as Escheresque disks, with most of their interior volume piled up near a spherical boundary surface. Semiclassical thinking about gravity and spacetime in some sense unifies AdS and its antipode, Stephen asserted. *It is as if the sign of λ has no real meaning in the realm of quantum gravity.*

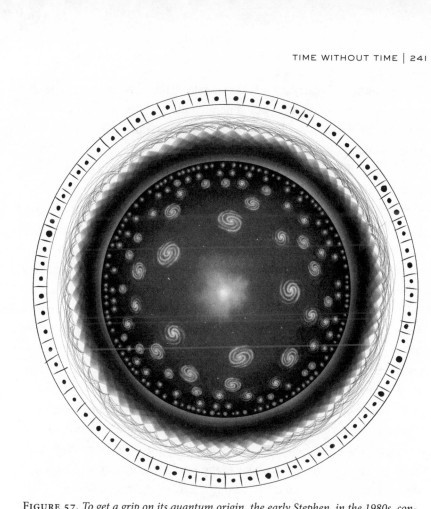

FIGURE 57. *To get a grip on its quantum origin, the early Stephen, in the 1980s, conceived of the universe in imaginary time. Thinking in imaginary time, all dimensions behave as directions of space, two of which are shown here. The origin of the universe lies at the center of the disk and it expands outward in the radial (imaginary-time) direction. The universe today corresponds to the circular boundary. However, the later Stephen went further, much further, and took us from imaginary time to no time at all. Exploiting the holographic properties of gravity, we came to envisage the boundary of the disk as a hologram made of entangled qubits from which the interior spacetime— our past history—projects down. Holographic cosmology builds in a top-down view in which the past is in some sense contingent on the present.*

This insight paved the way for a holographic *dS–QFT duality*. Much as for AdS, the circular boundary surface in figure 57 provides a natural home for a holographic description of an expanding universe. Given the similarities, the dual fields and particles living in this surface may actually share quite a few properties with those of AdS holograms.[13] Physicists are working hard to understand how, by adjusting the nuts and bolts in holographic surface worlds, one can generate either a lifeless AdS space or a universe that inflates and brings forth galaxies and life. *The universe may have a boundary after all,* Stephen quipped.

The chief difference between holograms that mirror AdS interiors and those of inflating universes lies in the nature of the extra dimension that pops out. In the former case the emergent direction is a curved dimension of space. It is the interior depth of AdS. In the case of an expanding universe, the time dimension is emergent. That is, *history itself is holographically encrypted.* And this may well come to be seen as the most mind-boggling entry of all in the holographic dictionary!

Now, this may sound outrageous. However, the notion that time and cosmological expansion are emergent qualities of the universe follows naturally from the sequence of insights that we have encountered on our journey. When Georges Lemaître first floated the idea of a quantum origin, he already pondered that time might be emergent: "Time would only begin to have a sensible meaning when the original quantum had been divided into a sufficient number of quanta."[14] Fifty years later, Jim and Stephen's no-boundary description substantiated Lemaître's intuition, with their idea that time morphs into space as we approach the beginning. The holographic incarnation of their theory illustrated in figure 57 takes us farther into timelessness still. By etching gravity and cosmological evolution into scores of quantum interactions confined to a three-dimensional surface, holography dispenses with a prior notion of time altogether. In a holographic universe, time would, in a sense, be illusory. This makes the original no-boundary proposal look rather conservative indeed.

What a remarkable journey this has been, from Newton's absolute time to time without time. Thinking about the passage of time as a holographic projection still feels unfamiliar though, even to theoretical physicists. I expect it will be many more years before physicists will be able to decipher the kind of holograms that encode the wobbly expansion histories of hesitating universes like ours. Countless intricate mathematical subtleties, highly interesting in their own right, will keep physicists occupied for a long time to come. We shouldn't expect holography one day soon requiring us to rewrite the standard cosmology textbooks. This is especially so because Einstein's geometric language works perfectly fine as a way to describe most of the large-scale universe. On the other hand, we can expect holography to become supremely important where Einstein's theory fails—inside black holes and, especially, at the big bang. After all, this is the nature—and power—of holographic dualities. A particularly exciting possibility is that the holographic underpinnings of the expansion may have been crucially important during inflation, and that subtle imprints of this in the microwave background fluctuations might be detectable with future gravitational-wave observations. Time will tell!

AT A CONCEPTUAL level, holography seals the top-down approach to cosmology. The central tenet of holographic cosmology—that the past projects from a web of entangled quantum particles that form a lower-dimensional hologram—*implies* a top-down view of the universe. If, as holographic cosmology posits, the surface of our observations is in some sense all there is, then this builds in the backward-in-time operation that is the hallmark of top-down cosmology. Holography tells us that there is an entity more basic than time—a hologram—from which the past emerges. The evolving and expanding universe would be output, not input in a holographic universe.

In Stephen's semiclassical thinking about quantum cosmology, the three pillars of the top-down triptych—histories, genesis, and observership—were only loosely entwined. While there was a close interplay among all three elements, they remained conceptually distinct entities. As a consequence, doubts lingered over whether the three ingredients could—or even had to—be truly merged, whether the top-down approach really was the foundational change Stephen proclaimed it was. But the architecture of holographic cosmology proves Hawking right. Holography tightens the top-down triptych into a unifying knot that constitutes a genuinely novel framework for prediction. First, by excising time from our list of fundamentals, it fuses dynamics with boundary conditions. Second, by placing holographic entanglement prior to spacetime, it integrates observership. Moreover, the math behind holographic cosmology encapsulates this synthesis in a single unified equation, a preliminary version of which can be discerned on the blackboard behind Stephen in the insert, plate 11. This puts top-down thinking on firmer ground indeed.

With its emphasis on entanglement, holography puts the capacity of systems to store and process information at the heart of this tightened triptych. Holography envisions that physical reality isn't just made up of real things, like particles of matter and radiation or even the field of space-time, but that it takes a far more abstract entity as well: quantum information. This breathes new life into another one of Wheeler's bold and seemingly far-fetched ideas. Wheeler too liked to think of physical reality as some kind of information-theoretic entity, an idea he dubbed "it from bit." He held the view that the physical world ultimately derives its existence from bits of information that form an irreducible kernel at the heart of reality. "Every physical object, every 'it,'" he wrote, "derives its significance from 'bits,' binary yes-or-no units of information."[15] Thirty years later, holography realizes Wheeler's vision with qubits, basic units of quantum information (and with a few layers of craziness still unwrapped). According to the dS–QFT duality, quantum information inscribed in an abstract timeless hologram of entangled qubits forms the thread that weaves reality. If you take away the entanglement on the boundary surface, your interior world just falls apart.

Unlike binary bits of ordinary information, which are zero or one, qubits consist of quantum particles that can be in a superposition of zero and one at the same time. When individual qubits interact, their possible states become entangled, each one's chances of zero and one hinging on those of the other. The entanglement means that if you measured some qubits, you could also learn something about their entangled partners, even when they are far away. Obviously, entangling in more and more qubits exponentially increases the number of simultaneous possibilities, and this is what makes quantum computers theoretically so powerful. The distributed manner in which quantum entanglement can store information helps to compensate for the fact that individual qubits are notoriously error-prone, which is the main challenge to the construction of quantum computers. The feeblest magnetic field or electromagnetic pulse can cause qubits to flip and derail calculations. So quantum engineers like to work with spatially distributed, entangled qubits and develop specialized schemes that build in redundancy, in order to protect quantum information even when individual qubits get corrupted. In fact, the efforts to design error-correcting codes that can cope with the daunting error rates of physical qubits is one of the major thrusts in the race to build a quantum computer.

It's an awesome twist really that in the meantime, in the aftermath of the holographic revolution sweeping theoretical physics, string theorists have begun to develop their own quantum error-correcting codes—to construct spacetime! In fact, the way an interior spacetime projects in holographic dualities rather resembles a highly efficient quantum error-correcting code. This might explain how spacetime acquires its intrinsic robustness, despite being woven out of such fragile quantum stuff. Some theorists have gone as far as suggesting that spacetime *is* a quantum code. They regard the lower-dimensional hologram as some sort of source code, operating on a huge network of interconnected quantum particles, processing information and, in this way, generating gravity and all other familiar physical phenomena. In their view the universe is a kind of quantum information processor, a vision that appears only a hair's breadth away from the idea that we live in a simulation.

Holography paints a universe that is being continually created. It is as if there is a code, operating on countless entangled qubits, that brings about physical reality, and this is what we perceive as the flow of time. In this sense holography places the true origin of the universe in the distant future, because only the far future would reveal the hologram in its full glory.

WHAT ABOUT THE distant past? How does a timeless cosmology conceive of the origins of time? Suppose that tomorrow theorists identify the hologram that corresponds to our expanding universe and we set out to read it, the AdS-QFT dictionary at hand, journeying back in time. What would we find all the way back at the bottom of spacetime?

One ventures into the past in holographic cosmology by taking something like a blurred viewpoint on the hologram. It is like zooming out. Remember that in Maldacena's duality, one moves deeper into the interior of AdS by considering larger scales in the surface hologram. Objects located at the very center of AdS are holographically encoded as long-range correlations across the entire hologram. Likewise, a hologram of an expanding universe inscribes the far past in qubits spanning huge distances in the surface world. We move farther into the past—toward the center of the disk in figure 57—by peeling away layer upon layer of information in the hologram, until we are left with a few distantly entangled qubits only. From a holographic point of view, the universe's earliest moments are definitely the spookiest action of all. In effect, eventually, one runs out of entangled bits. This, then, would be the origin of time.[16]

THE EARLY (BOTTOM-UP) Hawking regarded the no-boundary proposal as a description of the creation of the universe from nothing. In those days, Stephen strove to give a fundamentally causal explanation of the universe's origin: why, not how. But holography advances a more radical interpretation of his theory. Holographic cosmology shows that what Stephen's time-turns-into-space transition is really trying to tell, is that physics itself fades away when we journey back into the big bang. The no-boundary hypothesis emerges from holography not so much as a law of the beginning but more as a beginning of law. What is left, then, of the age-old question of the ultimate cause of the big bang? It would seem to evaporate. Not the laws as such but their capacity to change and transmute would have the final word.

This notion of cosmogenesis as a truly limiting frontier that emerges from holographic cosmology has far-reaching implications for multiverse cosmology. There is no evidence of a mosaic of island universes in any of the holograms physicists have ever devised. On the contrary, the interior wave functions that are holographically encoded appear to encompass only a very small patch of the string landscape: *Holographic cosmology excises the multiverse like Ockham's razor,* Stephen concluded.* In the last few years of his life, he was adamant that the multiverse frenzy had been an artifact of "classical, bottom-up thinking tying itself into knots."

The multiverse is in many respects the cosmological analogue of the (semi)classical theory of black holes. The latter fails to identify that there is an upper bound on the amount of information that black holes can

---

* Stephen was referring to the principle in philosophy, often attributed to the thirteenth-century English philosopher William of Ockham, that "entities are not to be multiplied without necessity."

store. Likewise, multiverse cosmology assumes that our cosmological theories can contain an arbitrarily large amount of information without affecting the cosmos they describe. But holographic cosmology paints a very different picture. The cosmic quilt of island universes stretching into all corners of the string theory landscape appears to be lost in uncertainty in holographic cosmology. Instead of an actual physical superstructure, that landscape may be better thought of as a mathematical realm that can inform physics but need not exist as such, somewhat like what the table of Mendeleev is to biology. *Asking what lies beyond our own universe, would be like asking which slit the electron goes through in the two-slit experiment,* Stephen put it. We are living in a patch of spacetime, surrounded by an ocean of uncertainty about which, well, we must remain silent.

Toward the end of our journey, I bumped into Andrei Linde at a conference and asked him how, twenty years on, he felt about the multiverse. To my surprise, Andrei said he thought that to get a grip on the multiverse, one would need to adopt a proper quantum view of the role of observers in cosmology. Had he thought that all along? Of course he hadn't. Science is what scientists do. We advance by exchanging ideas, through argument and reason, based on available evidence and abstraction. It required the deep paradoxes of the multiverse to bring the limitations of the conventional paradigm in physics, going back to Newton, more sharply into focus. Andrei's work inspired us to find the crack that got the quantum in. Without the extraordinarily frustrating and difficult conundrums that multiverse theory saddled us with, we would probably still be out there, looking for the view from nowhere, lost and confused in the nothingness beyond space and time.

STEPHEN LIFTED THE veil on our challenge to the multiverse in November 2016, during a cosmology session at the seat of the Pontifical Academy of Sciences in commemoration of Georges Lemaître. I wasn't surprised when Stephen told me he was keen to take part in this convocation. After all, it was at the Vatican in 1981 where he had first proposed that the universe had no boundary. Having turned the universe's history inside out and upside down, he must have felt he owed the academy an update on the subject that had been so close to his heart for so many years.

It would be Stephen's last journey abroad, and it turned into a difficult expedition. Hawking's doctors no longer let him fly on any of his friends' jets. Instead, he had to be transported by an air ambulance, and not just any air ambulance but one from a particular Swiss company. This was expensive, and with the Pontifical Academy of Sciences short on cash, we had to find a way to budget this on our research grants that cover only economy fares. When his doctors still refused to sign off on the trip, Stephen informed them he was scheduled to meet with Pope Francis. In the end they didn't want to quarrel with the prospect of a divine audience (although, coming from Stephen, they might have been suspicious of this claim), and Stephen was able to fly to Rome.

And so it happened that, thirty-five years after he first spoke at the Vatican, Stephen sat again in the Pontifical Academy's temple behind St. Peter's Basilica, where he explained that there is a dual description of the cosmos, a completely different and profoundly counterintuitive way of looking at reality, in which the expansion of space—and indeed time itself—is a manifestly emergent phenomenon, stitched out of a myriad of quantum threads forming a timeless world lying in a lower-dimensional surface. *The universe may have a boundary after all.*[17]

A FEW WEEKS before his passing I visited Stephen at home. He was nearly locked in at that point but surrounded by the best possible care. He knew he would die soon. In his study at Wordsworth Grove, we reconnected one last time. *I was never a fan of the multiverse,* he laboriously composed, as if I hadn't realized. *Time for a new book . . . include holography* were his last words to me—a final homework problem. I believe Stephen felt that the new holographic perspective on the universe would eventually make our top-down approach to cosmology seem evident and that we would one day wonder how we could have missed this for so long.

It was snowing heavily, as if nature provided Stephen with a blanket for his final voyage. Walking back to college, through Maltings Lane and over Coe Fen, across the Cam and past the Mill before circling around the old DAMTP, I reflected on our journey. In our search for the ultimate underpinnings of reality we had, in some kind of curious loop of interconnections, been led back to our own observations. "We are a way for the universe to know itself," Carl Sagan famously said. But it seems to me that in a quantum universe—our universe—we are getting to know ourselves. Top-down cosmology, whether in its holographic form or not, is grounded in our relationship with the universe. There is a subtly human aspect to it. And I have had the strong feeling, on many occasions, that the switch from a God's-eye to a worm's-eye view of the cosmos has felt like a home-coming to Stephen Hawking indeed.

FIGURE 58. *At work on our final theory with Jim Hartle at Cook's Branch, Texas.*

CHAPTER 8

AT HOME IN THE UNIVERSE

The fabric of existence weaves itself whole.

—CHARLES IVES

IN 1963, HANNAH ARENDT TOOK PART IN THE ESSAY CONTEST OF A Symposium on Space organized by the editors of *Great Ideas Today*. This was soon after the first human expeditions into space and at the time of NASA's plans to launch the Apollo 11 lunar mission. Arendt was asked whether "man's conquest of space had increased or diminished his stature." The seemingly obvious answer would be that of course, yes, it had increased man's stature. Arendt, however, didn't subscribe to this view.

In her essay "The Conquest of Space and the Stature of Man," she reflects on how science and technology transform what it means to be human.[1] Central to her concept of humanism is the idea of freedom. The freedom to act and be meaningful, she held, is what enables us to be human.[2] Arendt went on to ponder whether human freedom is threatened when we increasingly acquire the know-how to redesign and control the world, from our physical environment and the living world to the nature of intelligence.

Born to a German-Jewish family in 1906 in Hannover, Hannah Arendt studied at Marburg University under Martin Heidegger, but, like Einstein, she was forced to flee Germany in 1933. It was a lesson firsthand at how human freedom and dignity can be curtailed. For the next eight years she lived in Paris, then emigrated to the United States in 1941, where she became part of a lively intellectual circle in New York. Later, in her coverage for *The New Yorker* of the war crimes trial of Adolf Eichmann in Jerusalem, she famously (or notoriously, in the view of others) argued that ordinary people become complacent actors in totalitarian systems because they cease to think freely (or even to think at all) and disengage from the world. She traced such excesses in the sociopolitical sphere to the social corrosion brought about by what she called *world alienation,* the loss of a sense of belonging to the world and of the recognition that we are all bound together, a sense that there is a oneness to humanity and the civic engagement this bond entails.

Arendt felt strongly that modern science and technology lay at the roots of man's estrangement from the world. As a matter of fact, she pointed to the central insight that sparked the modern scientific revolution—the idea that the world is objective—as the main culprit. Ever since its inception, modern science has sought a higher truth, governed by rational and universal laws. In their pursuit of these, scientists succumbed to what Arendt termed *earth alienation* (not to be confused with world alienation), the search for an Archimedean standpoint from which it is hoped that such an objective understanding can be leveraged.

Her central thesis was that this position is antithetical to humanism. Of course, the scientific approach has been phenomenally successful, in both theoretical and practical terms, and its benefits to humanity are undeniable. But the flight from our earthly roots that is the hallmark of modern science has also led to a chasm between our human goals and the supposedly objective workings of nature. Over the span of nearly five centuries, Arendt held, this chasm has grown and increasingly challenged human nature, altering the societal fabric and, slowly but steadily, turning earth alienation, the view from nowhere intrinsic to much of science, into world alienation, a widespread disengagement from the world.

In her essay, Arendt puts her finger on this conundrum at the heart of modern science and argues that this will ultimately prove to be a self-defeating paradigm. Interestingly, in support of her thesis she quotes the quantum pioneer Werner Heisenberg, who said that "man in his hunt for objective reality suddenly discovered that he always confronts himself alone."[3] Heisenberg was referring here to the key role of the observer in quantum theory, the fact that the very questions one asks affect how reality is manifested. The instrumentalist interpretation of the theory that he and Bohr advanced and that typified the early quantum era generated a deep epistemological puzzle. Physicists were told to "shut up and calculate" and not to worry about the ontology of quantum theory. But Arendt did exactly that and pointedly noted that it was as if the sciences, with the advent of quantum theory, did what the humanities knew all along but could never demonstrate, namely, that humanists were right to be concerned about the stature of man in the new scientific world.

For Arendt, the launch of Sputnik, an event "second in importance to no other," epitomized the evolution toward an entirely artificial world, a "technotope" subject to human mastery and control. In her essay, she writes: "The astronaut, shot into outer space and imprisoned in his instrument-ridden capsule where each actual physical encounter with his surroundings would spell immediate death, might well be taken as the symbolic incarnation of Heisenberg's man—the man who will be the less likely ever to meet anything but himself and man-made things the more ardently he wishes to eliminate all anthropocentric considerations from his encounter with the non-human world around him."

To Arendt, this pursuit of science and technology stripped from all anthropomorphic elements and humanistic concerns was fundamentally flawed. Be it the conquest of space in the hopes of geoengineering another planet, or the search for the philosophers' stone in biotechnology—or indeed the quest for a final theory in theoretical physics—to her these were acts of rebellion against our human condition as dwellers on this planet:

> Man will necessarily lose his advantage. All he can find is the Archimedean point with respect to the earth but once arrived there and having acquired this absolute power over his earthly habitat, he would need a new Archimedean point, and so ad infinitum. Man can only get lost in the immensity of the universe, for the only true Archimedean point would be the absolute void behind the universe.

Arendt argued that if we begin to look down upon the world and our activities as if we are outside it, if we begin to lever ourselves, then our actions will ultimately lose their deeper meaning. This is because we would begin to see the Earth as an object like any other and no longer as our home. Our activities, from online shopping to our scientific practice, would be reduced to mere data points that can be analyzed with the same methods we use to study particle collisions or the behavior of rats in the lab. Our pride in what we can do would dissolve into some kind of mutation of the human race, transforming us from subjects of Earth to mere objects. If it should ever reach this point, Arendt concludes her essay, "the stature of man would not simply be lowered by all standards we know of but will have been destroyed." That is, we would lose our freedom. We would cease to be human.

This is the paradox. In our attempt to find the ultimate truth and absolute control over our existence as humans on Earth, we risk ending up smaller, not larger.

At the core of Arendt's argument lay the idea that science and technology could only truly add to the stature of man in as much as we desire to be at home in the universe. "Earth is the very quintessence of the human condition," she held. Whatever we find out about or do to the world are human discoveries and endeavors. No matter how abstract and imaginative our thoughts, or how far-reaching their impact, our theories and our actions remain inextricably interwoven with our human, earthly conditions. And this is why Arendt pleaded for a practice of science and a vision of the technotope grounded in our humanness:

> The new worldview that may conceivably grow out of modern science is likely to be once more geocentric and anthropomorphic. Not in the old sense of the earth being the center of the universe and of man being at the zenith of existence. But it would be geocentric in the sense that the earth, and not a point outside the universe, is the center and home of humanity. And it would be anthropomorphic in the sense that man would count his own finiteness among the elementary conditions under which his scientific efforts are possible at all.

This is where Hannah meets Stephen. That is, the later, top-down Stephen. Hawking's final theory frees cosmology from its Platonic straitjacket. It brings the physical laws in a sense back home. Adopting an inside-out perspective on the universe, the theory is rooted in what Arendt would call our earthly conditions. This isn't a mere abstruse academic matter, for a physical cosmology that recognizes the finitude inherent in our worm's-eye perspective on the cosmos will, in time, reorient the very scientific agenda. Indeed, if the past can be a guide, we may hope that Hawking's final theory can be the kernel of a new scientific and human worldview where man's knowledge and creativity will once again revolve around their common center.

COSMOLOGY MAY WELL be the one field of science in which the validity of Hannah Arendt's concerns is beyond doubt. Of course we are within the universe! Nevertheless, ever since Newton, cosmologists have striven to reason from a point outside it, and by the end of the twentieth century, multiverse speculation had turned earth alienation into *universe alienation*. Confounded by the biophilic nature of the supposedly objective laws and lost in the multiverse, cosmologists ended up smaller, not larger, much as Arendt foresaw.

What Arendt did not anticipate, I believe, was that Heisenberg's new quantum theory, in which "man confronts himself alone," also contained the seeds for cosmology to reinvent itself. In this book I have argued that a genuine quantum outlook on the universe counters the relentless alienating forces of modern science and lets one build cosmology anew from an interior viewpoint—the essence of Hawking's final theory.

IN A QUANTUM universe, a tangible past and future emerge out of a haze of possibilities by means of a continual process of questioning and observing. This observership, the interactive process at the heart of quantum theory that transforms what might be into what does happen, constantly draws the universe more firmly into existence. Observers—in this quantum sense—acquire a sort of creative role in cosmic affairs that imbues cosmology with a delicate subjective touch. Observership also introduces a subtle backward-in-time element into cosmological theory, for it is as if the act of observation today retroactively fixes the outcome of the big bang "back then." This is why Stephen referred to his final theory as top-down cosmology; we read the fundamentals of the history of the universe backward—from the top down.

By integrating observership *within* its architecture, yet without assigning a privileged role to life, top-down cosmology steers clear of both the danger of being "lost in math," articulated by Arendt, and the pitfalls of the anthropic principle. Somewhat prosaically, one might say that Stephen's final theory conceives of man neither as a godlike figure hovering above the universe, nor as a helpless victim of evolution in the margins of reality, but as nothing more or less than himself. Having wrestled with the anthropic principle for much of his career, Stephen was evidently pleased with this outcome. Top-down cosmology turns the riddle of the universe's apparent design in a sense upside down. It embodies the view that down at the quantum level, the universe engineers its own biofriendliness. Life and the universe are in some way a mutual fit, according to the theory, because, in a deeper sense, they come into existence together.

In effect, I venture to claim that this view captures the true spirit of the Copernican revolution. When Copernicus put the sun at the center, he realized all too well that from then on one would need to take the motion of Earth around the sun into account in order to interpret astronomical observations correctly. The Copernican revolution did not pretend that our position in the universe is irrelevant, only that it isn't privileged. Five centuries on, top-down cosmology returns to these roots, and I like to think Arendt would have been pleased.

That being said, Hawking's final theory did not come about from a sudden sympathy for one or another philosophical position. If anything, Stephen tried to refrain from adopting one. He felt Einstein, with his static universe and his reluctance to embrace quantum theory, had been guided too much by his philosophical prejudices, and he strove to avoid making the same mistakes. We developed our top-down approach primarily in an attempt to untangle the paradoxes of the multiverse and find a better cosmological theory. In retrospect this endeavor turned out to be rather productive, philosophically.

THE DISCOVERY IN the late 1920s that the universe has a history is one of the most startling discoveries of all time. For almost a century we have been studying that history against the stable background of immutable laws of nature. But the essence of the theory that Stephen and I put forward is that this approach fails to convey the depth and the scope of what Lemaître found out about the universe. The quantum cosmology we propose reads the universe's history from within and as one that includes, in its earliest stages, the genealogy of the physical laws. In our view it is not the laws as such but their capacity to change that is fundamental. In this way top-down cosmology completes the conceptual revolution in our thinking about the universe that Lemaître initiated.[4]

To uncover the essence of what lies hidden in the earliest quantum stages, one must peel away the many layers of complexity that separate us from the universe's birth. This can be done by tracing the universe back in time. When one finally gets to the big bang, a deeper level of evolution opens up in which the laws of physics themselves change. One discovers a sort of meta-evolution, a stage where the rules and principles of physical evolution co-evolve with the universe they govern.

This meta-evolution has a Darwinian flavor, with its interplay of variation and selection playing out in the primeval environment of the early universe. Variation enters because random quantum jumps cause frequent small excursions from deterministic behavior and occasional larger ones. Selection enters because some of these excursions, especially the larger ones, can be amplified and frozen in the form of new rules that help shape the subsequent evolution. The interaction between these two competing forces in the furnace of the hot big bang produces a branching process— somewhat analogous to how biological species emerge billions of years later—in which dimensions, forces, and particle species first diversify and then acquire their effective form when the universe expands and cools to ten billion degrees or so. The randomness involved in these transitions means that, just like the Darwinian evolution, the outcome of this truly ancient layer of cosmic evolution can only be understood *ex post facto.*

Of course, connecting the dots to piece together the tree of physical laws will continue to challenge us for the foreseeable future. With only scant fossil records of the earliest moments and with most of the universe's contents dark and mysterious, the cosmogenesis has proven extremely difficult to decipher. But advances in telescope technology continue to expand our senses. From exquisite observations of the microwave background radiation to ingenious searches for particles of dark matter and bursts of gravitational waves, physicists all over the world are gearing up to unlock that remote era harboring our deepest roots.

Now, if the effective laws of physics are fossil relics of an ancient evolution, then, ontologically speaking, we should probably regard these on equal footing with law-like characteristics of other levels of evolution. Stronger still, one might argue that in the grand frame of quantum cosmology, it appears that there isn't the slightest ontological difference between the fact that Christian religions dominated Western Europe at the dawn of the modern scientific age and, say, the value of the anomalous magnetic moment of the electron in the Standard Model of particle physics. They are both frozen accidents, just at widely different levels of complexity.

STEPHEN'S NO-BOUNDARY MODEL of the beginning—conceived from the top down!—is key to realize the fundamentally historical perspective on physics and cosmology that I have advocated, a view of physics that includes the genesis of the laws. The no-boundary hypothesis predicts that if we trace the primordial universe as far back in time as we possibly can, its structural properties continue to evaporate and transmute and that this extends, ultimately, to time itself. Time would initially have been melded with space into something like a higher-dimensional sphere, closing the universe into nothingness. This led the early Hawking, still reasoning in a causal bottom-up fashion, to proclaim that the universe was created from nothing. But Hawking's final theory offers a radically different interpretation of this closure of spacetime at the big bang. The later Hawking held that this nothingness at the beginning is nothing like the emptiness of a vacuum, out of which universes may or may not be born, but a much more profound, epistemic horizon involving no space, no time, and, crucially, no physical laws. "The origin of time" in Stephen's final theory is the limit of what can be said about our past, not just the beginning of all that is. This view is especially borne out by the holographic form of the theory where the dimension of time and hence the basic notion of evolution, the epitome of reductionist concepts, are seen as emergent qualities of the universe. From a holographic viewpoint, going back in time is like taking an increasingly fuzzy look at the hologram. One quite literally sheds more and more of the information that it encodes until, well, one runs out of qubits. That would be the beginning.

It is a striking property of top-down cosmology that it has a built-in mechanism that limits what we can say about the world. It is as if a proper quantum take on the cosmos guards us from wanting to know too much. And this is important, for it is precisely the closure of our past in Hawking's final theory, and the fundamental recognition of a certain finitude that this closure enforces, that prevents us from getting mired down in the paradoxes of the multiverse. The multiverse evaporates like snow before the sun in quantum cosmology. Top-down cosmology strips the variegated cosmic quilt of most of its colors, but this reduction, curiously, boosts the theory's predictive reach. So, as anticipated by Hannah Arendt in her incisive analysis, by jettisoning the Archimedean standpoint, cosmological theory emerges larger, not smaller. To quote Wittgenstein at the end of his famous *Tractatus:* "Whereof one cannot speak, thereof one must be silent." The power of a quantum outlook on the cosmos is that it gives us the mathematical tools to be silent indeed.

The upshot is a profound revision of what we understand that cosmology can ultimately find out about the world. The early Hawking (as well as the early author) sought a deeper understanding of the universe's apparent design in the physical conditions at the origin of time. He (we) assumed there was a fundamental causal explanation hidden deep in the math governing the big bang that would determine "why the universe is the way it is," as Stephen so often put it. That is, we assumed there was a final theory that superseded the physical universe—or multiverse. Having turned cosmology upside down and inside out, the later Hawking claimed that his earlier alter-ego had erred. Our top-down perspective reverses the hierarchy between laws and reality in physics. It leads to a new philosophy of physics that rejects the idea that the universe is a machine governed by unconditional laws with a prior existence and replaces it with the view that the universe is a kind of self-organizing entity, in which all sorts of emergent patterns appear, the most general of which we call the laws of physics. One might say that in top-down cosmology, the laws serve the universe, not the universe the laws. The theory holds that if there is an answer to the great question of existence, it is to be found within this world, not in a structure of absolutes beyond it.

I summarized the broad principles behind the top-down approach in the interconnected triptych that I sketched in figure 43. This scheme generalizes the conventional paradigm of physics in which the three pillars—histories, genesis, and observership—weren't entwined but conceived as separate, disjunct entities, each with its own status. The triptych amounts to a novel framework for prediction in which the inductive process of constructing laws about the universe is described within it and in which, as a consequence, our physical theories are seen as one possibility among many. The top-down view exudes that, most honestly, the laws of physics are properties of the universe that we induce from our collective data, compressed into computational algorithms,[5] not manifestations of some external truth. The succession of physical theories is understood as the identification of ever more general patterns, encompassing an ever-larger number of interconnected empirical phenomena. Of course this progression greatly enhances the predictive power and the utility of physical theory, but it is a different reading altogether to say that this puts us on a path toward a final theory that would be unique, independent of its construction, and independent of our data. It is a basic observation indeed that there are always a great many theories that fit a finite set of data, just as there are many curves that interpolate between a finite set of points. Likewise, the top-down approach to cosmology should lead us to suspect that we'll find a succession of theories all the way down, but not an endpoint. In a way Stephen's final theory states there isn't a final theory.

Freed from any claims on absolute truth, top-down cosmology provides space for a multitude of spheres of thoughts, from art to science, each serving different purposes and spurring complementary insights. If our top-down thinking does contain the seeds of a new worldview, then it is a thoroughly pluralistic one. Notions of time and law-like patterns are seen to emerge in a way that is contingent on the questions we ask and grounded in the complexity of the universe around us. When the later Hawking outlined our post-Platonic cosmology at the Vatican, in November 2016, there were no more battles with God or the pope to be fought. Quite on the contrary, Stephen found a strong and moving resonance with Pope Francis in their shared goal of protecting our common home in the cosmos for the benefit of humanity today and tomorrow.

WE LEARN FROM quantum cosmology that biological evolution and cosmological evolution aren't fundamentally separate phenomena but two vastly different levels of one giant evolutionary tree. Biological evolution is concerned with branches up in a high-complexity realm, whereas cosmology deals with low-complexity layers, with the astrophysical, geological, and chemical levels filling in the ranges in between. And even though every level has its own specificity, its own language, the universal wave function weaves them all together.[6] The "higgledy-piggledy" way in which the tree of physical laws emerged in the early universe shows that the broad principles of Darwinism, that quintessential biological scheme, reach all the way down into the deepest level of evolution we can imagine. Quantum cosmology in some sense bridges that gnawing conceptual chasm that for eons has separated biology and physics. It tells us that Darwin's sketch of the tree of life and Lemaître's sketch of a hesitating universe (see insert, plates 4 and 3, respectively) are deeply connected, representing two stages of a single overarching historical process.

Such an extraordinary arc reveals a profound and powerful unity in nature. Vastly different levels of evolution are merging into an interconnected whole, with correlations interlinking them. The leitmotif all along our journey, the stunning fitness for life of the effective laws of physics, is arguably the preeminent example of a correlation across multiple levels of complexity. We can now begin to understand at a deeper level how we, a twiglet on the tree of life, together with all other species on our planet, interconnect with the physical universe around us and piece together what it is that breathes life in the cosmos. In effect Charles Darwin, in his farsightedness, may already have anticipated this development. In 1882 in a letter to George Wallich, Darwin wrote: "The principle of continuity renders it probable that the principle of life will hereafter be shown to be part, or consequence, of some general law encompassing all of Nature." We may finally be on the verge of realizing Darwin's vision.

Nevertheless, many physicists, especially theorists (who tend to have strong opinions on the deeper roots of the laws of nature) would still rather believe that there is a final theory out there, hovering above and beyond physical reality—a rock-solid foundation of the tower of science at the center of existence. This mindset didn't escape Stephen.[7] "Some people will be very disappointed if in the end there is no ultimate theory," he noted. But he continued: "I used to belong to that camp. I'm now glad that our search for understanding will never come to an end, and that we will always have the challenge of new discovery. Without it, we would stagnate." In quintessential Hawking fashion, Stephen was ready to move on, eager to embark on an exciting journey of post-Platonic discovery.

LIKE DARWIN, STEPHEN felt there was grandeur in this view. And it is a tremendously exciting prospect indeed! If all scientific laws are emergent laws, the "fundamental" laws of physics included, we are on the brink of discovering a much broader view of nature. As a matter of fact, these insights tie in with recent developments in a range of scientific disciplines. By jettisoning the idea that we are looking for the unique set of rules, science in several corners is branching out from studying "what is" to "what may be."

In the information sciences, AI and machine learning are creating new forms of computing and intelligence, some with the capacity to evolve and even acquire an element of intuition—human or otherwise. Bioengineering unlocks novel evolutionary pathways based on different genetic codes and even proteins. Genetic editing techniques such as CRISPR,* for example, allow geneticists to modify a cell's DNA in precise and targeted ways, designing life-forms with shapes or capabilities that do not exist in "natural nature." These range from genius mice to long-living worms, and perhaps, one day, genius long-lived humans, or rather post-humans. Meanwhile, quantum engineers make new forms of matter that display the weirdness of microscopic quantum entanglement on the macroscopic scales of everyday life. Some of these materials may even holographically encrypt new theories of gravity and black holes, or even toy expanding universes, whose evolution is encoded in algorithmic operations on a large number of interconnected quantum bits.

These are far-reaching developments. Instead of merely discovering the laws of nature by studying phenomena that exist, scientists are starting to envisage hypothetical laws and then engineer systems in which they emerge. The old goal to find *the* nature of intelligence, or *the* theory of everything, may soon be seen as a relic of an outdated and overly limited worldview. In a recent *Quanta Magazine* article, Robbert Dijkgraaf, former director of Princeton's Institute for Advanced Study, writes, "What we

---

* Clustered regularly interspaced short palindromic repeats

used to call 'nature,' is only the tiniest fraction of a vastly larger landscape out there, waiting to be unlocked."[8]

Moreover, these developments reinforce each other, and it is at their intersection that we may well find the most sweeping consequences. In 2020, a deep-learning program called AlphaFold, developed by Google's AI branch DeepMind, trained itself to determine the three-dimensional folded shape of proteins from their sequence of amino acids, solving one of the big open challenges in the field of molecular biology. In the next few years, machine learning algorithms will be searching for new particles in the petabytes of data produced in the LHC at CERN and for patterns of gravitational waves in the noisy vibrations picked up at LIGO. In time we should expect such deep-learning programs to dive with us into the mathematical structures underpinning our physical theories and, who knows, to reimagine the basic language of physics.

So, by embracing the realm of "what could be," we have arrived on the cusp of a whole new chapter in the age of modern science. In the twentieth century, scientists identified the elementary building blocks of nature: particles, atoms, and molecules are the constituents of all matter; genes, proteins, and cells are the components of life; bits, codes, and networked systems underpin intelligence and information. In this century we will begin to engineer new realities with their own laws by connecting these components in novel ways. Of course the rest of the natural world has been doing this for over thirteen billion years of cosmological expansion and almost four billion years of biological evolution on Earth. But as Dijkgraaf eloquently describes, it has only explored the tiniest fraction of all possible designs. The number of genes that can be conceived mathematically is stupendous, far larger even than the number of microstates of a typical black hole, but only a small sliver of these have been realized in life on Earth. Likewise, the range of physical forces and particles that can be fabricated in string theory is enormous. But the expansion of the early universe produced only this specific set. So, across the entire spectrum of complexity, from fundamental physics to intelligence, the manifold of possible realities is immensely larger than what natural evolution has so far produced. The twenty-first century is that critical period in history in which we are beginning to unlock this immense realm.

This transition signifies the dawn of a new era, the first of its kind in the history of Earth, and perhaps even of the cosmos, in which a species attempts to reconfigure and transcend the biosphere it has evolved in. Echoing Hannah Arendt, from merely undergoing evolution, we are transitioning toward engineering it and, with it, our humanness.

On the one hand, this is a time of great promise. The sheer latitude of pathways opening up is truly fantastic compared to anything we have ever experienced before. In some branches of the future, our choices today will act as a springboard for unimaginable innovation and post-human flourishing. In these futures, the human era will represent a remarkable passage between the first nearly four billion years of agonizingly slow, Darwinian evolution and the next untold years of evolution driven by technological and intelligent design—both here on Earth and far beyond.

But this is also a dangerously precarious time. Human-made existential risks, from the proliferation of nuclear war and global heating to advances in biotech and artificial intelligence, now far outweigh those naturally occurring. Britain's Astronomer Royal Sir Martin Rees has estimated that taking all risks into account, there is only a 50 percent chance that we will get through to 2100 without a disastrous setback. Oxford's Future of Humanity Institute puts the existential risk for humanity this century at around one in six. Hence there are countless future paths, not just an improbable branch here or there, where we may descend into chaos and even vanish, leaving no more than a minor footnote in cosmic history.

We have only one firm data point related to our prospects: No alien civilizations appear to have explored a substantial fraction of the stellar systems in our cosmic neighborhood. Hence among the billions of stars in our local past light cone, none appear to have evolved into a large-scale ecosystem with the level of technology that we may soon reach. The physical laws are remarkably fit for life, yet there is no evidence of anybody out there. We haven't been able to tune in to the alien ham radio beaming extraterrestrial poetry, nor have we seen mysterious astroengineering projects across the sky. On the contrary, we have had great success at explaining the behavior of stellar systems, our galaxy, and the entire observable universe on the basis of a single set of natural physical laws. Wondering about this paradox, the Italian physicist Enrico Fermi, in the summer of 1950, famously asked, "Where is everybody?" Fermi's point was that the lack of evidence for extraterrestrial civilizations, given such biophilic conditions, suggests that there is a serious obstacle somewhere on the road of evolution from ordinary dead matter to the advanced technotope we may soon be. Do the principal bottlenecks lie in our past or in our future, or both? If the evolutionary steps in our past are so incredibly improbable that complex life-forms are rare in the universe, then the main bottleneck would almost certainly be behind us. But Fermi had a nagging feeling that the roadblock may lie in the one transition that separates our current civilization from being able to spread into the cosmos: We may not be able to survive the world we've created. Some more insight in this would be helpful to come to a measure of collective foresight as we carve out a future.[9] As a matter of fact, Stephen shared Fermi's feeling, saying at one point: "We only have to look at ourselves to see how intelligent life might develop into something we wouldn't want to meet."

This leads us to the question: What kind of future do we envisage for our planet and our species? Will post-human life flourish and expand into the cosmos? Taking a quantum view, the myriad of paths forking off into the future are in a sense already out there, as a landscape of possibilities. Some futures may even appear rather plausible. We should learn from the past, though, that chance constantly interferes, leading history to take unexpected twists and turns. The accidental behavior of a bat in Wuhan sometime in 2019 is just one example. Yet we can outline steps to avoid the precipice by acquiring a clear global vision of the kind of future we aspire to and, despite the uncertainties, modeling on a somewhat quantitative basis how it might function. A major responsibility in this will fall to the community of scientists and scholars to act as a societal think tank and ensure that their research is integrated and directed to the common good—from bioengineering to machine learning and quantum technology.

For we cannot simply wait and hope for the best. If humanity can't even collectively envision a future it aspires to, we can scarcely hope to attain anything remotely resembling it. There is no manual out there for us to consult and no foundations, not even, I have argued, at the bottom of the laws of physics, to cushion any failure. If mankind does not write its own script, no one will do it for us. We can either let evolution take its blind course, lowering the stature of mankind to that of a large-scale ant colony, collectivized and monitored, deprived of all freedom; or we can recognize that our fate lies in our own hands and, step by step, mold that fate into a coordinated vision of what constitutes a future that could prove Fermi's pessimism wrong.

At this critical point in history, when we take our first steps in nature's shoes, it will be more important than ever to remember Hannah Arendt's message that we are riders on planet Earth, not gods acting from the heavens. We are agents within a constantly changing universe. We are evolution. We need to find a way toward a planetary consciousness, in order to alleviate Arendt's world alienation and move toward a perspective on the world that redraws our relations with one another and with the rest of the biosphere in a way that values the future. Only by treasuring that we are stewards of planet Earth, and the finitude that comes with it, will we be able to avoid humanity pitting its many powers against itself.

BY REVOKING THE view from nowhere, Stephen's final theory offers a powerful kernel of hope. Our journey into the big bang was about OUR origins, not merely the origins of the universe starting with the big bang. That was such a key part of it. Like Einstein, Stephen thought that humanity's long-term future would ultimately depend on how well we understood our deepest roots. This is what drove him to study the big bang. His final theory of the universe is more than just a scientific cosmology. It is a cosmology in the humanistic sense, in which the universe is seen as our home—albeit a big one—and its physics rooted in our relationship with it. Hawking's cosmological finale bridges Isaac Newton's mathematical rigor with Charles Darwin's profound insight that in a deeper sense, we are one. It is indeed proper that Stephen's ashes are now buried in between the graves of Newton and Darwin in the nave of London's Westminster Abbey.

Throughout my journey with Stephen, I came to know him as someone who longed for all of us to embrace more of a cosmic perspective on our existence and think in terms of deep time. His final theory is like a budding seed that has the potential to grow into a new worldview thoroughly based on science and at the same time grounded in our humanity. Obviously, the arc from quantum cosmology to a moral universe is extremely long and fragile. But so is the arc laid out by Arendt, from Galileo observing the moon to the high-tech society today.

Stephen firmly believed that the courage of our questions and the depth of our answers would allow us to navigate planet Earth safely and wisely into the future. The story of his life in which he found, after his terrifying diagnosis with ALS, the will to love, to have children, to experience the world in all of its dimensions, and to grasp the universe, inspired millions and will remain a powerful metaphor for what humanity can achieve. His parting message, beamed into space during a memorial service on June 15, 2018, in Westminster Abbey, encapsulates it all: "When we see the earth from space we see ourselves as a whole; we see the unity and not the divisions. It is such a simple image, with a compelling message: one planet, one human race. Our only boundaries are the way we see ourselves. We must become global citizens. Let us work together to make that future a place we want to visit."

From Stephen Hawking we can learn to love the world so much that we aspire to reimagine it, and never to give up. To be truly human. Though he was nearly immobile, Stephen was the freest man I have known.

ACKNOWLEDGMENTS

MY JOURNEY WITH STEPHEN HAWKING WOULD NOT HAVE BEEN POSSI-
ble without the help of many colleagues and friends along the way.

Thanks to Adrian Ottewill and Peter Hogan from Dublin, Ireland, who in 1996 put me on the train to Cambridge, UK. Sincere thanks to Neil Turok, whose captivating classes in this mecca of theoretical cosmology encouraged me to knock on Stephen's door. And to my fellow PhD students in Hawking and Turok's orbit, including Christophe Galfard, Harvey Reall, James Sparks, and Toby Wiseman, for their comradeship.

"You should go away as far as possible," Stephen said when I graduated, and I did. Many thanks to Steve Giddings, David Gross, Jim Hartle, Gary Horowitz, Don Marolf, Mark Srednicki, and the late Joe Polchinski, for creating such an extraordinary, stimulating research environment at UC Santa Barbara in these exciting early days of string cosmology.

Around this time, Stephen bonded with George Mitchell. Sincere thanks to the Mitchell family for creating a wonderful refuge at their Cook's Branch Conservancy where Stephen was able to work. Special thanks also to the International Solvay Institutes in Brussels, to their president Jean-Marie Solvay and their longtime director Marc Henneaux, and to the Institutes' mater familias, Madame Marie-Claude Solvay, whose perspicuous reminiscences about Oppenheimer, Feynman, or Lemaître make the history of twentieth-century physics come alive. The Solvays' warmth and generosity meant that the Institutes became so much more than a scientific haven on our journey.

Over the years a great many conversations with numerous colleagues have profoundly influenced my thinking on the origin of time. A special thanks for this to Dio Anninos, Nikolay Bobev, Frederik Denef, Gary Gibbons, Jonathan Halliwell, Ted Jacobson, Oliver Janssen, Matt Kleban, Jean-Luc Lehners, Andrei Linde, Juan Maldacena, Don Page, Alexei Starobinsky, Thomas Van Riet, Alex Vilenkin, and, again, Gary Horowitz, Joe Polchinski, Mark Srednicki, and Neil Turok. Thanks also to the European Research Council and the Research Foundation Flanders, for support of the technical research that underpins the broader cosmological theory that I develop in this book.

Of course, working with Stephen would have been impossible without his support teams, his sequence of graduate assistants and personal secretaries, especially Jon Wood and Judith Croasdell, and the many caregivers and nurses whose professional and creative care, tinkering, and scheduling kept Spaceship Hawking flying well beyond mission duration.

Profound thanks are due to Jim Hartle, our *compagnon de route* on this exhilarating voyage, whose seemingly innate quantum take on the universe was always a shining beacon on the horizon, and to Tom Dedeurwaerdere, my invaluable sounding board and source of inspiration.

I am indebted to the Centre for Theoretical Cosmology at Cambridge and its benefactors, and to Trinity College, for a visiting fellowship at a critical fork in the road. To Martin Rees, and the Pontifical Academy of Sciences, who facilitated the dissemination of an early version of Stephen's final cosmological theory.

Sincere and special thanks to Lucy Hawking for her gentle and courageous steering, especially in the difficult later stages, when Stephen's final days drew near and the idea of narrating our journey was born. The first few lines of this book were written around the kitchen table at Wordsworth Grove.

I have aimed to frame our collaborative endeavors against the broader historical development of both relativistic and quantum cosmology. For illuminating discussions on this history, I thank the late John Barrow, Gary Gibbons, Dominique Lambert, Malcolm Longair, and Jim Peebles. Special thanks also to Frans Cerulus for sharing, at age ninety-five, his still vivid personal memories of Abbé Georges Lemaître. To Liliane Moens and Véronique Fillieux for their invaluable assistance in navigating the rich Lemaître Archives at the Université catholique de Louvain; and to Graham Farmelo, for an enlightening discussion on Hawking's early scientific and personal life.

The commitment of my close colleagues at KU Leuven, Nikolay Bobev, Toine Van Proeyen, and Thomas Van Riet, to a vibrant research group at the Institute for Theoretical Physics also created a stimulating writing environment despite the challenges of COVID-19 lockdown. Thanks also to my wider circle of colleagues in Leuven and in the Low Countries, from the visionaries nurturing a precious academic environment where scientific writing for a more general readership has a home, to the heroes endeavoring to test our most advanced cosmological theories. A special word of thanks to Robbert Dijkgraaf for, perhaps unwittingly, providing much inspiration and encouragement.

To Demis Hassabis for an eye-opening conversation on what the future(s) of cosmology might be—and might mean—in the era of AI. To the playwright Thomas Ryckewaert, who bravely took that spectrum of ideas (and the author) to the stage. To H.R.H. Queen Mathilde of Belgium, for her delightful visit to the exhibition *To the Edge of Time* in Leuven. And to my co-curator Hannah Redler Hawes for venturing enthusiastically into the vast open space between science and art, adding in the process a slight artistic touch to this oeuvre.

I also thank and congratulate the archivists of the VRT, the Flemish public broadcasting corporation. The ink was barely dry on this manuscript when they found the long-lost recording of a 1964 interview with Georges Lemaître offering striking support for the intellectual arc from him to the later Hawking that I establish in this book.

I thank Aïsha De Grauwe, who masterfully turned my sketches into the images that illustrate the text, and Georges Ellis, Roger Penrose, and James Wheeler for their kind help with some of the older images. I also want to express my appreciation to the curators of Hawking's office at the London Science Museum and of the Paul A. M. Dirac Papers at Florida State University.

For good advice and guidance throughout this book project, many thanks are due to my literary agents, Max Brockman and Russell Weinberger. And to Hilary Redmon, my fine editor at Random House, for her sharp editorial insights and continued encouragement, as well as to Miriam Khanukaev, for navigating the manuscript through production.

Finally, heartfelt thanks to Nathalie and to our children, Salomé, Ayla, Noah, and Raphael, for creating such a wonderful and loving home throughout my journey.

ILLUSTRATION CREDITS

Figure 26: © photo by Anna N. Zytkow

Figure 32: personal archives of Professor Andrei Linde

Figure 39: © Maximilien Brice/CERN

Figure 40: © photograph by Paul Ehrenfest, courtesy of AIP Emilio Segrè Visual Archives

Figure 44: © *The New York Times*/Belga image

Figure 47: reproduced from John A. Wheeler, "Frontiers of Time," in *Problems in the Foundations of Physics, Proceedings of the International School of Physics "Enrico Fermi,"* ed. G. Toraldo di Francia (Amsterdam; New York: North-Holland Pub. Co., 1979/KB-National Library)

Figure 55: © M.C. Escher's *Circle Limit IV* © 2022 The M.C. Escher Company, The Netherlands. All rights reserved. www.mcescher.com

Figure 56: reproduced from John A. Wheeler, "Geons," *Physical Review* 97 (1955): 511–36

Figure 58: © photo by Anna N. Zytkow

COLOR INSERT

Plate 1: © Georges Lemaître Archives, Université catholique de Louvain, Louvain-la-Neuve, BE 4006 FG LEM 836

Plate 2: first published in Algemeen Handelsblad, July 9, 1930, "AFA FC WdS 248," Leiden Observatory Papers

Plate 3: © Georges Lemaître Archives, Université catholique de Louvain, Louvain-la-Neuve, BE 4006 FG LEM 704

Plate 4: public domain

Plate 5: © *The New York Times Magazine.* First published on Feb. 19, 1933.

Plate 6: © Succession Brâncuși—all rights reserved (Adagp)/Centre Pompidou, MNAM-CCI /Dist. RMN-GP

Plate 7: first published in Thomas Wright, *An Original Theory of the Universe* (1750)

Plate 8: M.C. Escher's "Oog" © The M.C. Escher Company—Baarn, The Netherlands. All rights reserved. www.mcescher.com

Plate 9: © ESA—European Space Agency/Planck Observatory

Plate 10: © Science Museum Group (UK)/Science & Society Picture Library

Plate 11: © Sarah M. Lee

BIBLIOGRAPHY

Arendt, Hannah. *The Human Condition*. Chicago: University of Chicago Press, 1958.

Barrow, John, and Frank Tipler. *The Anthropic Cosmological Principle*. Oxford: Oxford University Press, 1986.

Carr, Bernard J., George F. R. Ellis, Gary W. Gibbons, James B. Hartle, Thomas Hertog, Roger Penrose, Malcolm J. Perry, and Kip S. Thorne. *Biographical Memoirs of Fellows of the Royal Society: Stephen William Hawking CH CBE, 8 January 1942–14 March 2018*. London: Royal Society, 2019.

Carroll, Sean. *The Big Picture: On the Origins of Life, Meaning, and the Universe Itself*. London: Oneworld, 2017.

Davies, Paul. *The Goldilocks Enigma: Why Is the Universe Just Right for Life?* London: Allen Lane, 2006.

Farmelo, Graham. *The Strangest Man: The Hidden Life of Paul Dirac, Mystic of the Atom*. New York: Basic Books, 2009.

Gell-Mann, Murray. *The Quark and the Jaguar*. New York: Freeman, 1997.

Greene, Brian. *The Fabric of the Cosmos*. New York: Alfred A. Knopf, 2004.

Greene, Brian. *The Hidden Reality: Parallel Universes and the Deep Laws of the Cosmos*. New York: Alfred A. Knopf, 2011.

Halpern, Paul. *The Quantum Labyrinth*. New York: Basic Books, 2018.

Hawking, Stephen. *A Brief History of Time: From the Big Bang to Black Holes*. New York: Bantam Books, 1988.

Hawking, Stephen, and Leonard Mlodinow. *The Grand Design*. New York: Bantam Books, 2010.

Lambert, Dominique. *The Atom of the Universe: The Life and Work of Georges Lemaître*. Kraków: Copernicus Center Press, 2011.

Nussbaumer, Harry, and Lydia Bieri. *Discovering the Expanding Universe.* Cambridge: Cambridge University Press, 2009.

Pais, Abraham. *"Subtle Is the Lord—": The Science and the Life of Albert Einstein.* Oxford: Oxford University Press, 1982.

Peebles, James. *Cosmology's Century: An Inside History of Our Modern Understanding of the Universe.* Princeton: Princeton University Press, 2020.

Pross, Addy. *What Is Life?* Oxford: Oxford University Press, 2012.

Rees, Martin. *If Science Is to Save Us.* Cambridge: Polity Press, 2022.

Rees, Martin, *Our Cosmic Habitat.* Princeton: Princeton University Press, 2001.

Rovelli, Carlo. *The First Scientist: Anaximander and His Legacy.* Translated by Marion Lignana Rosenberg. Yardley, Pa: Westholme, 2011.

Smolin, Lee. *The Trouble with Physics: The Rise of String Theory, the Fall of Science and What Comes Next.* Boston: Mariner Books, 2007.

Susskind, Leonard. *The Cosmic Landscape: String Theory and the Illusion of Intelligent Design.* New York: Little, Brown, 2006.

Susskind, Leonard. *The Black Hole War.* New York: Little, Brown, 2008.

Turok, Neil. *The Universe Within: From Quantum to Cosmos.* Toronto: House of Anansi Press, 2012.

Weinberg, Steven. *To Explain the World: The Discovery of Modern Science.* New York: Harper, 2015.

Wheeler, John Archibald, and Kenneth Ford. *Geons, Black Holes, and Quantum Foam: A Life in Physics.* London: Norton, 1998.

NOTES

PREFACE

1. After Stephen's passing, this blackboard was acquired for the nation by the London Science Museum Group, together with other memorabilia of Hawking's Cambridge office. The scribbles aren't Stephen's, it turns out, but are from participants of the monthslong conference, including Hawking's co-organizer and postdoc at the time, Martin Roček, whose face can be seen slightly drawn on the right roughly in the middle.
2. Christopher B. Collins and Stephen W. Hawking, "Why Is the Universe Isotropic?" *Astrophysical Journal* 180 (1973): 317–34.
3. Stephen occasionally lent his voice, a process whereby someone would draw up a statement that would then be run through his speech software and broadcast to the outside world. Those around him, however, could easily distinguish fake Hawking phrases from real ones—the latter standing out in concision, clarity, and his trademark sense of humor. While this practice was necessary for several reasons, it was also unfortunate, for it meant that the public image of Hawking became gradually separated from the real person.

CHAPTER 1: A PARADOX

1. Fred Hoyle, "The Universe: Past and Present Reflections," *Annual Review of Astronomy and Astrophysics* 20 (1982): 1–36.
2. Steven Weinberg, "Anthropic Bound on the Cosmological Constant," *Physical Review Letters* 59 (1987): 2607.
3. Paul Davies, *The Goldilocks Enigma: Why Is the Universe Just Right for Life?* (London: Allen Lane, 2006), 3.

4. This fragment has come down to us through Simplicius of Cilicia, who quotes it in his commentary on Aristotle's *Physics*.

5. Galileo Galilei, *Il Saggiatore* (Rome: Appresso Giacomo Mascardi, 1623).

6. Phrase attributed to François Arago.

7. Paul Dirac, as quoted in Graham Farmelo, *The Strangest Man: The Hidden Life of Paul Dirac, Mystic of the Atom* (New York: Basic Books, 2009), 435.

8. William Paley, *Natural Theology; or, Evidences of the Existence and Attributes of the Deity, Collected from the Appearances of Nature* (London: Printed for R. Faulder, 1802).

9. Charles Darwin, *On the Origin of Species*, manuscript, 1859.

10. Stephen Jay Gould, *Wonderful Life: The Burgess Shale and the Nature of History* (New York: Norton, 1989).

11. Charles Darwin, as quoted in Charles Henshaw Ward, *Charles Darwin: The Man and His Warfare* (Indianapolis: Bobbs-Merrill, 1927), 297.

12. Leonard Susskind, *The Cosmic Landscape: String Theory and the Illusion of Intelligent Design* (New York: Little, Brown, 2006).

13. Despite what the name suggests, neither Carter nor anyone else thinks of the anthropic principle as being concerned with humankind specifically, but rather with the conditions for life more generally. A detailed review of the idea can be found in John Barrow and Frank Tipler, *The Anthropic Cosmological Principle* (Oxford: Oxford University Press, 1986).

14. Andrei Linde, "Universe, Life, Consciousness" (lecture, Physics and Cosmology Group of the "Science and Spiritual Quest" program of the Center for Theology and the Natural Sciences [CTNS], Berkeley, Calif., 1998).

15. Steven Weinberg, Living in the Multiverse, delivered at the Symposium "Expectations of a Final Theory" at Trinity College, Cambridge, September 2005, and published in *Universe or Multiverse?*, ed. B. Carr (Cambridge: Cambridge University Press, 2007).

16. Nima Arkani-Hamed, "Prospects for Contact of String Theory with Experiments" (lecture, Strings 2019, Flagey, Brussels, July 9–13, 2019).

17. Hawking repeated this in his lecture "Cosmology from the Top Down" (lecture, Davis Meeting on Cosmic Inflation, University of California, Davis, March 22–25, 2003).

18. In *The Structure of Scientific Revolutions*, the American philosopher of science Thomas Kuhn explained that paradigm shifts arise when the reigning paradigm under which established science operates is rendered incompatible with new phenomena. One may wonder what exactly were the "new phenomena" that cropped up and triggered the call for change in cosmology around the turn of the twenty-first century. Chiefly among these, I believe, were the astronomical observations in the late 1990s of the acceler-

ated expansion. These conspired with the new theoretical insights in string theory to exemplify the accidental nature of the biophilic laws.

19. Hawking, working with his student Bernard Carr in the mid-1970s, speculated about the existence of small black holes that would have formed in the in the wake of the hot big bang. Such *primordial black holes* would be hotter and radiate faster. In fact, those of about 10^{15} grams—that is, with the mass of a mountain, but the size of a proton—would be exploding in the present epoch of the universe. To Stephen's great disappointment, no such explosions have been seen.

CHAPTER 2: DAY WITHOUT YESTERDAY

1. Georges Lemaître, "Rencontres avec Einstein," in *Revue des Questions scientifiques* (Bruxelles: Société scientifique de Bruxelles, January 20, 1958), 129.

2. Georges Lemaître, in "Univers et Atome," his last public lecture, delivered in 1963 to an audience of former Louvain students. This phrasing is somewhat stronger than the way he usually described his position, which no doubt reflects a certain frustration with the attitude of his opponents. An in-depth account of Lemaître's (somewhat evolving) views on the relation between science and religion, including an analysis of this lecture, is given by Dominique Lambert in *L'itinéraire spirituel de Georges Lemaître* (Bruxelles: Lessius, 2007).

3. Thomson was ennobled in 1892 as 1st Baron Kelvin of Largs. The title refers to the river Kelvin, which flows near his laboratory at the University of Glasgow. Today we know Lord Kelvin primarily because his name was given to the absolute scale of temperature. Kelvin determined that the value of absolute zero temperature is approximately −273.15 degrees Celsius. In an epic undertaking, he also laid the first transatlantic telegraph cable, between Ireland and Newfoundland. Here Kelvin is quoted in Lord Kelvin, "Nineteenth Century Clouds over the Dynamical Theory of Heat and Light," *Philosophical Magazine* 6, no. 2 (1901): 1–40.

4. Hermann Minkowski, "Raum und Zeit" (lecture, 80th General Meeting of the Society of Natural Scientists and Physicians, Cologne, September 1908).

5. Quoted in Abraham Pais, *"Subtle Is the Lord—": The Science and the Life of Albert Einstein* (Oxford: Oxford University Press, 1982).

6. The language of warped geometry that Einstein employed had been developed in the nineteenth century by mathematicians such as Carl Friedrich Gauss and Bernhard Riemann, who realized that the usual rules of geometry that many of us learn at school, like the famous theorem named after

Pythagoras, or the theorem that the angles of a triangle add up to 180 degrees, don't work on curved surfaces. For example, on an orange (or on the earth's surface), the angles of a triangle add up to more than 180 degrees. Before Gauss and Riemann, curved surfaces had always been thought of as embedded in normal three-dimensional Euclidean space. But Gauss showed that the geometrical properties of two-dimensional curved surfaces, like notions of straight lines and angles, can be defined intrinsically, without referring to anything outside them. This opened up the way for Riemann to imagine that in a similar fashion, a three-dimensional space could be curved and differ from a Euclidean space. Einstein imagined exactly this and went one step further, describing the physical world in terms of a four-dimensional warped geometry of spacetime. Curved spacetime obeys the rules of non-Euclidean geometry in four dimensions, without having to invoke anything outside or beyond it. Physically, this means, e.g., that the universe doesn't need to exist or expand in some sort of bigger box.

7. John Archibald Wheeler and Kenneth Ford, *Geons, Black Holes, and Quantum Foam: A Life in Physics* (London: Norton, 1998), 235.

8. Pais, *"Subtle Is the Lord."*

9. Special cable to *The New York Times,* November 10, 1919.

10. This wasn't the first time this radius popped up in physics. Already in the 1700s, using Newtonian mechanics, John Michell and Pierre-Simon Laplace found that a spherical mass M compressed within this radius would have an escape velocity equal to the speed of light. Such hypothetical objects would not be able to radiate particles of light and can be seen as precursors of black holes.

11. See, e.g., Georges Lemaître, "L'univers en expansion," *Annales de la Société Scientifique de Bruxelles* A53 (1933): 51–85. Available in English translation as "The Expanding Universe," *General Relativity and Gravitation* 29, no. 55 (1997): 641–80.

12. During most of its life, a normal star will support itself against its own gravity by thermal pressure generated by nuclear fusion converting hydrogen into helium. Eventually, however, a star will exhaust its nuclear fuel and contract. If the star is not too massive to begin with, the pressure from the repulsion between electrons (or between neutrons and protons) will eventually halt the collapse and the star will settle down as a white dwarf (or a neutron star). However, the Indian-American astrophysicist Subrahmanyan Chandrasekhar earned his Nobel Prize by showing, in 1930, that white dwarfs have a maximum mass. Next, in 1939, Robert Oppenheimer and George Volkoff showed that neutron stars too have a maximum mass. The upshot is that there is no known state of matter that could halt the gravita-

tional collapse of sufficiently massive stars, which are thought to continue to contract, producing a black hole.

13. Roger Penrose, "Gravitational Collapse: The Role of General Relativity," *La Rivista Del Nuovo Cimento* 1 (1969): 252–76.

14. Roger Penrose, "Gravitational Collapse and Space-time Singularities," *Physical Review Letters* 14, no. 3 (1965): 57–59.

15. The Einstein equation on page 42 contains a quantity, $8\pi G/c^4$, multiplying the mass and energy content of the matter on the right-hand side. The numerical value of this quantity is extremely small, which means that one needs an enormous amount of mass or energy to deform the spacetime field residing on the left-hand side of the equation ever so slightly. To give an idea, the mass of the entire planet Earth deforms the shape of space in its neighborhood, compared to normal Euclidean space, by something of the order of 10^{-9}.

16. Einstein, letter to Willem de Sitter, March 12, 1917, in *Collected Papers*, vol. 8, eds. Albert Einstein, Martin J. Klein, and John J. Stachel (Princeton University Press, 1998): Doc. 311.

17. For a more detailed account of the history of the discovery of the expansion I recommend Harry Nussbaumer and Lydia Bieri's *Discovering the Expanding Universe* (Cambridge: Cambridge University Press, 2009).

18. I warmly recommend the biography of Georges Lemaître, *The Atom of the Universe*, by Dominique Lambert (Kraków: Copernicus Center Press, 2015).

19. Lemaître is quoting Saint Thomas Aquinas here, who said, "Nothing exists in the intellect that was not previously in the senses."

20. Georges Lemaître, "L'Etrangeté de l'Univers," a lecture he gave to the Circolo di Roma in 1960, reprinted in *Pontificiae Academiae Scientiarum Scripta Varia* 36 (1972): 239.

21. Cepheids are pulsating stars whose luminosity rises and falls over periods ranging from months down to merely a day. Henrietta Leavitt, one of the first female astronomers of the modern era, had earlier noticed a curious relation between the pulsating period of Cepheids and their luminosity; Cepheids that were dimmer had shorter periods. This meant one could use observations of the periodic variations in brightness of Cepheids to measure distances in cosmology. And so it happened that Cepheids became astronomers' first reliable yardstick to distant objects in the universe, used by Hubble with great ingenuity to estimate the distances to nebulae.

22. The Lowell Observatory had been founded in 1894 by Percival Lowell to study the mysterious "channels" on Mars. In 1930 it was the place where Pluto was discovered.

23. The spectrum of light is the way it is distributed over different colors. A

shift in the spectrum of light from an astronomical object can be established by comparing the wavelength of an identifiable feature in the spectrum to the wavelength of the same characteristic feature when it is measured in a laboratory on Earth.

24. Vesto M. Slipher, "Nebulae," *Proceedings of the American Philosophical Society* 56 (1917): 403–9.

25. His paper was written in French and published in the rather obscure *Annales de la Société Scientifique de Bruxelles* (Série A. 47 [1927]: 49–59). Its title, "Un univers homogène de masse constante et de rayon croissant, rendant compte de la vitesse radiale des nébuleuses extragalactiques" ("A homogeneous universe of constant mass and increasing radius, accounting for the radial velocities of extragalactic nebulae"), leaves no doubt about Lemaître's intentions. In effect, Lemaître slightly modified it during his final editing of the manuscript, changing "variant" to "croissant," probably to strengthen the connection between his model and the astronomical observations, suggesting galaxies are moving away from us.

26. Lambert, *Atom of the Universe.*

27. Because of the large uncertainties on the distances, Lemaître divided the mean value of the velocities by the mean value of the distances in the sample of galaxies for which Hubble had published estimates of distances. Taking the mean helped average out the large uncertainty in each individual distance measurement.

28. Seeking to continue his conversation with Einstein, Lemaître got into the taxi that was to take Einstein to the laboratory of Auguste Piccard, his former student in Berlin. During the taxi ride, Lemaître brought up the subject of the observed recession of the nebulae and that this provided some evidence for a universe in expansion. However, he came away, according to his recollection, with the impression that Einstein was neither informed nor interested in the latest astronomical observations.

29. Friedmann's breadth of expertise extended from purely mathematical work on relativity to dramatic high-altitude balloon flights to investigate the effects of altitude on the human body. He held the world's ballooning altitude record for a period in 1925, ascending to a height of 7,400 meters (24,278 feet), higher than the tallest mountain in Russia. He died a few months later, apparently from typhus fever, at the age of thirty-seven.

30. Like Einstein, Lemaître had a strong philosophical preference for a spatially finite universe.

31. In 2018 the International Astronomical Union adopted a resolution that the relationship should be referred to as the Hubble-Lemaître law.

32. On the basis of improved observations of twenty-four galaxies, Hubble ob-

tained a value for the proportionality constant H in the velocity–distance relation on page 55 of 513 km/sec for every three million light-years of distance—not all that different from the value found earlier by Lemaître. Hubble and Humason interpreted their observations in terms of an ordinary Doppler shift.

33. Einstein, letter to Tolman, 1931, in Albert Einstein Archives, Archivnummer 23-030.

34. Arthur Stanley Eddington, *The Expanding Universe* (Cambridge: Cambridge University Press, 1933), 24.

35. Georges Lemaître, Evolution of the expanding universe, Proceedings of the National Academy of Sciences, 20, 12–17.

36. Einstein, letter to Lemaître, 1947, in Archives Georges Lemaître, Université catholique de Louvain, Louvain-la-Neuve, A4006.

37. Hubble and Humason's redshift observations only take us back a few million light-years. Hence their measurements determined the rate of expansion in relatively recent cosmic times but said nothing about how that rate has evolved throughout the history of the universe. In the Golden 1990s, spectral observations of bright supernova explosions, which could be seen out to billions of light-years away, made it possible to reconstruct the universe's course of expansion billions of years back in time. This revealed that our universe transitioned from a slowing expansion toward acceleration about five billion years ago.

38. Georges Lemaître, *Discussion sur l'évolution de l'univers,* (Paris: Gauthier-Villars, 1933), p 15–22.

39. Lemaître belonged to a new breed of mathematical astronomers convinced that the future of astronomy would involve pure analysis as well as computer programming. His computational research followed very closely the progress in computing technology. Early on in the 1920s he assisted Vannevar Bush at MIT by testing the differential analyzer on the Störmer problem. Later he moved his calculations of orbits of cosmic rays from tables of logarithms to hand-cranked adders, then on to electric desk machines and mechanically automated accounting machines, and he finally realized his dream when Douglas Hartree gave him access to the vacuum tube computer in development at Cambridge University in the 1950s.

40. Arthur S. Eddington, "The End of the World: from the Standpoint of Mathematical Physics," *Nature* 127, no. 2130 (March 21, 1931): 447–53.

41. Lemaître, *Revue des Questions scientifiques.*

42. Lemaître, "L'univers en expansion."

43. Georges Lemaître, "The Beginning of the World from the Point of View of Quantum Theory," *Nature* 127, no. 2130 (May 9, 1931): 706.

44. P.A.M. Dirac, in "The Relation Between Mathematics and Physics," a lecture he delivered on February 6, 1939, on presentation of the James Scott Prize. Published in the *Proceedings of the Royal Society of Edinburgh* 59 (1938–39, Part II): 122–29.

45. Fred Hoyle, "The Universe: Past and Present Reflections," *Annual Review of Astronomy and Astrophysics* 20 (1982): 1–36.

46. Fred Hoyle, *The Origin of the Universe and the Origin of Religion* (Wakefield, R.I.: Moyer Bell, 1993).

47. Many more anecdotes about his colorful life can be found in Gamow's autobiography, *My World Line: An Informal Autobiography* (New York: Viking Press, 1970).

48. Heavier chemical elements like carbon were fused much later through nuclear fusion inside stars. Even heavier elements, beyond iron, formed later still, either in the sudden heat of supernovae or in the violent mergers of neutron stars. These and other processes forged the chemically rich environment of the present-day universe. In effect, the most exotic elements of all are being fused today in the laboratories of physicists on Earth (and perhaps elsewhere).

49. Lambert, *Atom of the Universe*.

50. Quoted in Duncan Aikman, "Lemaitre Follows Two Paths to Truth," *The New York Times Magazine*, February 19, 1933 (see insert, plate 5).

51. Georges Lemaître, "The Primaeval Atom Hypothesis and the Problem of the Clusters of Galaxies," in *La structure et l'evolution de l'univers: onzieme conseil de physique tenu a l'Universite de Bruxelles du 9 au 13 juin 1958*, ed. R. Stoops (Bruxelles: Institut International de Physique Solvay, 1958): 1–30. Isaiah's notion of *Deus Absconditus*, the Hidden God, was a constant theme in the background of Lemaître's thinking. For example, the manuscript of his big bang manifesto in *Nature* in 1931 contains a short paragraph at the end—crossed out before publication—in which he writes, "I think that everyone who believes in a supreme being supporting every being and every acting, believes also that God is essentially hidden and may be glad to see how present physics provides a veil hiding the creation."

52. Lemaître, "The Primaeval Atom Hypothesis and the Problem of the Clusters of Galaxies."

CHAPTER 3: COSMOGENESIS

1. Stephen Hawking, *My Brief History* (New York: Bantam Books, 2013), 29.

2. Pressure came, for example, from surveys of radio sources, later known as quasars, which showed that these sources were distributed fairly uniformly

across the sky. This meant that they were probably outside our galaxy. But there were too many faint sources, indicating that their density had been higher in the distant past, not what one expects in an unchanging steady-state universe.

3. Like Penrose, Stephen identified a point of no return, namely the formation of an anti-trapped surface, from where light rays radiated in all directions diverge. Stephen showed that if there was once an anti-trapped surface, then there must have been a singularity a little farther back in time.

4. George F.R. Ellis, "Relativistic Cosmology," in *Proceedings of the International School of Physics "Enrico Fermi," Course 47: General Relativity and Cosmology,* ed. R. K. Sachs (New York and London: Academic Press, 1971), 104–82.

5. As quoted in *General Relativity and Gravitation: A Centennial Perspective,* A. Ashtekar, B. Berger, J. Isenberg, M. Maccallum, eds. (Cambridge: Cambridge University Press, 2015), 19.

6. Hendrik A. Lorentz, "La théorie du rayonnement et les quanta,": in *Proceedings of the First Solvay Council, Oct 30–Nov 3, 1911,* eds. P. Langevin and M. de Broglie (Paris: Gauthier-Villars, 1912), 6–9.

7. Heisenberg's uncertainty principle goes hand in hand with Planck's quantum hypothesis. Imagine you want to measure the position of a particle. To do so, you have to look at the particle, for example, by shining light on it. To measure the position more precisely, you can use light of a shorter wavelength. By Planck's quantum hypothesis, however, one has to use at least one quantum of light. This quantum will slightly disturb the particle, changing its velocity in a way that cannot be predicted. The shorter the wavelength, the higher the energy of a single quantum of light, and the larger the resulting uncertainty in the particle's velocity. Heisenberg's uncertainty principle quantifies this, stipulating that the uncertainty in the position of a particle times the uncertainty in its momentum can never be smaller than a certain quantity called Planck's constant, denoted by h. The value of Planck's constant can be determined experimentally. It is one of the fundamental constants of nature, alongside the speed of light, c, and Newton's gravitational constant, G, both of which appear in the Einstein equation on page 42. Planck's quantum constant, by contrast, is conspicuously absent from this classical (as opposed to quantum) equation!

8. Schrödinger's description of particles in terms of waves of probability also explains the early quantum experiments with atoms. Take, for example, an electron orbiting an atomic nucleus. If we regard the electron as a wavelike entity, then only for certain orbits will the length of the orbit correspond to a whole number of wavelengths of the electron. For these orbits the wave

crest would be in the same position each time around, so the waves would add up and reinforce one another. These are precisely Bohr's quantized orbits.

9. Erwin Schrödinger, *Science and Humanism: Physics in Our Time* (Cambridge: Cambridge University Press, 1951), 25.

10. For a colorful account of the scientific and personal interactions of Richard Feynman and John Wheeler, I warmly recommend Paul Halpern, *The Quantum Labyrinth* (New York: Basic Books, 2018).

11. Freeman J. Dyson, referencing Feynman in a statement in 1980, as quoted in Nick Herbert, *Quantum Reality: Beyond the New Physics* (Garden City, N.Y.: Anchor Press, 1987).

12. Imagine one cheats by stationing an apparatus near one of the slits to verify which path the electron *really* takes. With the extra detector in place, you will indeed see each electron passing through one slit or the other. However, you will also find that the interference pattern on the screen disappears. This is because with the new apparatus in place, we are asking a different question, thereby selecting a different set of histories. By adding the new apparatus we ask, "Which path did the electron take?" To answer this, Feynman's sum-over-histories scheme instructs us to add up all electron paths that pass through a given slit. Obviously this yields the total probability to pass through that slit, namely 50 percent. But by forcing the electron to reveal that information, we have also eliminated all histories passing through the other slit, and hence the possibility of interference between both sets of trajectories on their way to the screen. The interference pattern emerges only if the experimenter makes no attempt to determine which slit any given electron went through.

13. James B. Hartle and S. W. Hawking, "Path-Integral Derivation of Black-Hole Radiance," *Physical Review D* 13 (1976): 2188–203.

14. It will in due time be possible to find out more on the genesis of the no-boundary hypothesis: the UCSB archives contain a large blue ring binder labeled "81–82 Wave Function" in which Jim Hartle meticulously kept his correspondence with Stephen during those two crucial years.

15. Jim Hartle, private communication.

16. The diameter of the cigar specifies the temperature of the black hole as measured by a distant observer. The larger the diameter of the cigar, the lower the hole's temperature. For a given mass, the diameter is specified in the Euclidean framework by requiring that the geometry be smooth at its tip, like a sphere and not like a cone. This is how the Euclidean geometry of a black hole encrypts its quantum behavior.

17. Gary W. Gibbons and S. W. Hawking, eds., *Euclidean Quantum Gravity* (Singapore; River Edge, N.J.: World Scientific, 1993), 74.

18. Sidney Coleman, "Why There Is Nothing Rather Than Something: A Theory of the Cosmological Constant," *Nuclear Physics B* 310, nos. 3–4 (1988): 643.

19. This sort of topical discussion meeting had been established by Monsignor Lemaître in the 1960s, when he was president of the Pontifical Academy of Sciences.

20. Allocution of His Holiness John Paul II, published in *Astrophysical Cosmology: Proceedings of the Study Week on Cosmology and Fundamental Physics*, eds. H. A. Brück, G. V. Coyne, and M. S. Longair (Città del Vaticano: Pontificia Academia Scientiarum: Distributed by Specola Vaticana, 1982).

CHAPTER 4: ASHES AND SMOKE

1. The hot big bang theory also predicts a cosmic neutrino background, or CNB, and even a cosmic graviton background. The CNB would, if observed, give a snapshot of the universe when it was mere seconds old.

2. Georges Lemaître, *L'hypothèse de l'atome primitif: Essai de cosmogonie* (Neuchâtel: Editions du Griffon, 1946).

3. Bernard J. Carr et al., *Biographical Memoirs of Fellows of the Royal Society: Stephen William Hawking CH CBE, 8 January 1942–14 March 2018* (London: Royal Society, 2019).

4. In Newton's theory, gravity arises purely from an object's mass and energy, but in general relativity, pressure too contributes to an object's gravity, to how it warps spacetime. What's more, unlike mass, which is always positive, pressure can also be negative. A familiar example of negative pressure is the inward pull you feel when you stretch a rubber band. In Einstein's theory, positive pressure, like positive mass, contributes positively to gravity, but negative pressure leads to repulsive gravity, or antigravity.

5. The main players behind these theoretical predictions included Gennady Chibisov, Viatcheslav Mukhanov, and Alexei Starobinsky, working in Russia, and, in the West, James Bardeen, Alan Guth, Stephen Hawking, So-Young Pi, Paul Steinhardt, and Michael Turner.

6. G. W. Gibbons, S. W. Hawking, and S.T.C. Siklos, eds., *The Very Early Universe: Proceedings of the Nuffield Workshop, Cambridge, 21 June to 9 July, 1982* (Cambridge; New York: Cambridge University Press, 1983).

7. One member of a pair of virtual particles has positive energy and the other has negative energy. The negative-energy particle cannot remain in exis-

tence in an ordinary spacetime but has to seek out its positive-energy part-ner and annihilate with it. However, a black hole contains negative energy states, so if the negative energy member of a virtual pair falls into a black hole, it can continue to exist without having to annihilate itself with its partner, which is thus free to escape. The negative energy of the particle falling in slightly lowers the mass of the black hole, which explains why Hawking radiation makes black holes shrink and eventually disappear.

8. The very first indications that the universe contains more matter than meets the eye go back to the 1930s, to Swiss astronomer Fritz Zwicky's ob-servations of clusters of galaxies. Zwicky observed that some galaxies orbit other galaxies with surprisingly high velocities. This meant there had to be much more matter than what could be accounted for by the visible stars to hold such clusters together. In the 1970s, U.S. astronomer Vera Rubin ob-served a similar effect in the outskirts of individual galaxies. Her observa-tions indicated that also the arms of spiral galaxies must be embedded in a cloud of dark matter that holds it all together.

9. Two teams of astronomers, the High-Z Supernova Project jointly led by Adam Riess and Brian Schmidt, and the Supernova Cosmology Project led by Saul Perlmutter, measured the brightness and the redshift of light from exploding stars called supernovae, which are so bright that they are visible even in distant galaxies. Because the intrinsic brightness of these super-novae is known, the researchers were able to use these stars as distance benchmarks deep into the universe. Combined with their redshift observa-tions, this enabled both teams to establish the Hubble-Lemaître law relating distances and recession velocities out to billions of light-years, thus recon-structing the course of expansion billions of years back in time. To their surprise, their measurements showed that the expansion of the universe had begun to speed up roughly five billion years ago, a discovery for which Perl-mutter, Riess, and Schmidt shared the 2011 Nobel Prize.

10. Doubts persist as to whether the current acceleration of the expansion is driven by a truly constant cosmological constant or whether a very slowly changing scalar field is involved, a sort of remnant inflaton field. In the former case the ratio of pressure and energy density would be exactly equal to -1, whereas in the latter case it would be greater than -1. This difference may not seem terribly important but it affects the rate of acceleration in the (very) long run and hence may alter the universe's ultimate fate. Efforts are under way to determine the value of this ratio as precisely as possible.

11. A small cloud has since appeared. Relatively local astronomical observa-tions, such as those of the spectra of supernovae, point to a rate of expan-sion of 73 km/sec for every megaparsec in distance. By contrast, the rate of

expansion deduced from observations of the cosmic microwave background with the aid of general relativity comes out at around 67 km/sec for every megaparsec. This discrepancy is known as the "Hubble tension" although it should really be termed the "Hubble–Lemaître tension." Cosmologists are fervently searching for an explanation. Could it be that this is general relativity's Mercury moment, that the theory needs to be tweaked somehow? Stay tuned!

12. Wave functions in ordinary quantum mechanics, without gravity, obey the Schrödinger equation, which prescribes how they evolve in time. Time is the sole entity in ordinary quantum mechanics that doesn't interfere with anything else. Without any problem, physicists calculate probabilities in quantum mechanics for observations at a precise, given moment in time. All of this is only possible however because ordinary quantum mechanics presumes that there is a fixed and definite spacetime background, in which wave functions of particles evolve. By contrast, in quantum cosmology spacetime itself is quantum mechanical and fluctuating. As a consequence, we no longer have anything available that can function as a universal clock. Hence it should be no surprise that time drops out in a quantum description of the universe as a whole. True, the wave function of the universe obeys an abstract version of the Schrödinger equation, first written down by John Wheeler and Bryce DeWitt, but this is not a dynamic law. It is more like a timeless constraint on the wave function in its totality.

13. S. W. Hawking and N. Turok, "Open Inflation without False Vacua," *Physics Letters* B 425 (1998): 25–32.

14. To the best of my knowledge, the idea of eternal inflation was first mentioned by Linde in his contribution "The New Inflationary Universe Scenario," in *The Very Early Universe: Proceedings of the Nuffield Workshop, Cambridge, 21 June to 9 July, 1982,* eds. G. W. Gibbons, S. W. Hawking, and S. T. C. Siklos (Cambridge; New York: Cambridge University Press, 1983), 205–49.

15. Linde, "Universe, Life, Consciousness."

16. You may wonder how eternal inflation and the multiverse get around Hawking's theorem that there must have been a singularity in the past. It doesn't quite do so, as Guth, Vilenkin, and Arvind Borde showed. The theory of eternal inflation merely pushes the singularity much farther back into the past, but doubts persist whether it can be truly eternal.

CHAPTER 5: LOST IN THE MULTIVERSE

1. The antiproton is the antiparticle of the proton. It has electric charge −1, as opposed to the +1 electric charge of the proton. Paul Dirac predicted the

existence of the antiproton in his 1933 Nobel Prize lecture on the basis of the equation that bears his name. The antiproton was first experimentally found in 1955 at the Bevatron accelerator in Berkeley. Nowadays antiprotons are routinely detected in cosmic rays.

2. The reason is that the Higgs boson should also mingle with heavier, yet-to-be-found particles. This mingling should inflate its mass and, with it, the mass of everything else. Yet this is not the case, a puzzle known as the hierarchy problem in particle physics: There is a clean hierarchy, a huge energy split, between the relatively low masses and energies of the elementary particles in the Standard Model and far higher energy scales in nature, up to the Planck scale, where physicists believe microscopic quantum gravity effects become important. Theorists have conjectured that an exotic symmetry, known as supersymmetry, may be responsible for keeping the Higgs boson light. Supersymmetry says that every matter particle has a partner exchange particle, so it effectively doubles the species of elementary particles. Now, this supersymmetric doubling is such that the various contributions to the Higgs mass would perfectly cancel out, thus keeping it light. However, the LHC has searched in vain for partner particles predicted by supersymmetry, leading some to doubt they exist.

3. Quoted in his Scott Prize lecture. In effect, Dirac had a specific suggestion. He had noticed that three different combinations of some of the constants of nature all combine to roughly the same extremely large number, 10^{39}. This can't be a coincidence, he reasoned, and he speculated that a deeper law interlinked these quantities. The radical part of Dirac's suggestion was that he used the present age of the universe as one of the "constants" in some of the combinations he considered. Of course the age of the universe changes over time, so by assigning a fundamental meaning to these numerical coincidences, he also forced one of the traditional constants of nature to change as time passes. Dirac sacrificed the oldest "constant," Newton's gravitational constant, G, which had to be inversely proportional to the age of the universe for his arithmetic to work out. This turned out to be wrong: In a universe where gravity weakens over time, the energy output of the sun would have been much larger in the not-too-distant past, making the oceans on Earth boil in the pre-Cambrian era, to such a degree that life as we know it would not have evolved.

4. The idea that extra dimensions of space might have something to do with the unification of forces goes back to the work of the German mathematician Theodor Kaluza and the Swedish physicist Oscar Klein in the 1920s. Kaluza found that the Einstein equation applied to universes with one time dimension and four space dimensions describes not only gravity in the fa-

miliar four-dimensional spacetime, but also Maxwell's equations of electromagnetism. Electromagnetism emerges in Kaluza's setup from ripples propagating through the fourth space dimension. Klein then suggested that this extra dimension could be perfectly hidden from our senses if it were very small. Taken together, Kaluza and Klein's scheme provided an early example of the unifying power of extra dimensions.

5. Leonard Susskind, "The Anthropic Landscape of String Theory," in *Universe or Multiverse?*, ed. B. Carr (Cambridge: Cambridge University Press, 2007), 247–66.

6. Furthermore, the cosmological constant can't be very negative either, because this would lead to an additional attractive pull that causes the (island) universe to collapse again into a big crunch before galaxies get a chance to form.

7. The reason is that if the (island) universe starts out after inflation with larger primordial density variations, then the growth process of large-scale structures can better withstand the outward push of a cosmological constant. This broadens the range of values of λ that are compatible with the existence of galaxies, over several orders of magnitude.

8. For example, consider two kinds of island universes in the cosmic landscape that are equally habitable but have different particles making up the dark matter (with the same total amount of dark matter). Imagine that in one universe the extra curled-up dimensions of string theory generate very heavy dark matter particles that are impossible to produce in particle accelerators on Earth, while the other universe has a light dark matter particle that should be detectable with the successor of the LHC. Should we expect to find a dark matter particle when we turn on the next particle collider? This is a perfectly reasonable question of exactly the kind that experimental particle physicists (not to mention the governments and the public supporting physics research) would want to know the answer to. Clearly the anthropic principle doesn't help; both types of islands are equally fine from an anthropic viewpoint. Instead, one would need a theoretical prior that weighs the relative likelihood of both types of islands without relying on anthropic random selection. We return to this in the next chapter, where I will argue that this is precisely what a proper quantum outlook on cosmology provides.

9. For an eloquent criticism of random selection in cosmology, see James B. Hartle and Mark Srednicki, "Are We Typical?," *Physical Review D* 75 (2007): 123523.

10. Some of the founding fathers of cosmology, too, realized that prior probabilities or notions of typicality were of little use when considering a unique

system. Contemplating the quantum origin of the universe, Lemaître said, "The splitting of the atom can have occurred in many ways. The one that really occurred might have been very improbable." Dirac made a similar point in a letter to Gamow, who had criticized Dirac's time-varying gravity theory of the formation of the solar system on the grounds that it required the sun to have an improbable history. Dirac countered that he agreed that the sun's trajectory was improbable in his theory, but that this kind of improbability doesn't matter. "If we consider all the stars that have planets, only a very small fraction of them will have passed through clouds of the right density. . . . However, provided there is one, it is sufficient to fit the facts. So there is no objection to assuming our sun has had a very unusual and improbable history."

11. Private communication.

CHAPTER 6: NO QUESTION? NO HISTORY!

1. The crux of the conversations featuring in the first part of this chapter have appeared in published form in S. W. Hawking and Thomas Hertog, "Populating the Landscape: A Top-Down Approach," *Physical Review D* 73 (2006): 123527, and in S. W. Hawking, "Cosmology from the Top Down," *Universe or Multiverse?*, ed. Bernard Carr (Cambridge: Cambridge University Press, 2007), 91–99. See also Amanda Gefter's report "Mr. Hawking's Flexiverse," *New Scientist* 189, no. 2548 (April 22, 2006): 28.

2. Copernicus made the case for a heliocentric model based on mathematical simplicity, not on better agreement with astronomical observations. The first versions of the Copernican model of the solar system presumed circular planetary orbits and made nearly the same predictions for the apparent motions of the sun and planets as the geocentric Ptolemaic model. The idea that planets don't move in circles but on ellipses, a major departure from thousands of years of thinking, was put forward by Johannes Kepler in 1609, in his *Astronomia Nova,* in an attempt to reconcile the new Copernican theory with improved astronomical data of Tycho Brahe, Kepler's predecessor in Prague. But even Kepler's refinements of the heliocentric model could be mimicked in the Ptolemaic system by adding a few more epicycles. The first observational evidence that decisively favored heliocentrism over the ancient Ptolemaic system came only with Galileo's telescopic observations. Galileo saw that Venus had phases, like those of the moon, which could not be explained by any stretch of the Ptolemaic theory.

3. As for Copernicus himself, if he was a revolutionary, he was a reluctant one. His *De Revolutionibus Orbium Coelestium* was delivered to the print-

ers in 1543, shortly before his death, and its initial impact was muted. Furthermore, as if to comfort his readers, Copernicus pointed out that in his heliocentric model, Earth is "almost" in the center, writing, "Although the earth is not at the center of the world, nevertheless the distance to the center is nothing compared to that of the fixed stars."

4. This phrasing was used in a very different context by Thomas Nagel, *The View from Nowhere* (Oxford: Clarendon Press, 1986).

5. Sheldon Glashow, "The Death of Science!?" in *The End of Science? Attack and Defense,* Richard J. Elvee, ed., (Lanham, Md.: University Press of America, 1992).

6. Hannah Arendt, *The Human Condition* (Chicago: University of Chicago Press, 1958).

7. Stephen made a similar statement in public around the same time, in his lecture "Gödel and the End of Physics" delivered at Strings 2002 in Cambridge (UK).

8. As a matter of fact, the Solvay Councils still exist today and continue to enjoy generous support from the Solvay family.

9. Otto Stern, quoted in Abraham Pais, *"Subtle Is the Lord—": The Science and the Life of Albert Einstein* (Oxford: Oxford University Press, 1982).

10. Albert Einstein, "Autobiographical Notes," in *Albert Einstein, Philosopher-Scientist,* ed. Paul Arthur Schilpp (Evanston, Ill.: Library of Living Philosophers, 1949).

11. Einstein, letter to Max Born, December 4, 1926, in *The Born-Einstein Letters,* A. Einstein, M. Born, and H. Born, (New York: Macmillan, 1971), 90.

12. Quoted in J.W.N. Sullivan, *The Limitations of Science* (New York: New American Library, 1949), 141.

13. Hugh Everett III, "The Many-Worlds Interpretation of Quantum Mechanics" (PhD diss., Princeton University, 1957).

14. Bruno de Finetti, *Theory of Probability,* vol. 1 (New York: John Wiley and Sons, 1974).

15. John A. Wheeler, "Assessment of Everett's 'Relative State' Formulation of Quantum Theory," *Reviews of Modern Physics* 29, no. 3 (1957): 463–65.

16. John A. Wheeler, "Genesis and Observership," in *Foundational Problems in the Special Sciences,* eds. Robert E. Butts and Jaakkob Hintikka (Dordrecht; Boston: D. Reidel, 1977).

17. John A. Wheeler, "Frontiers of Time," in *Problems in the Foundations of Physics, Proceedings of the International School of Physics "Enrico Fermi,"* ed. G. Toraldo di Francia (Amsterdam; New York: North-Holland Pub. Co., 1979), 1–222.

18. Wheeler, "Frontiers of Time."

19. I regard our paper—S. W. Hawking and Thomas Hertog, "Populating the Landscape: A Top-Down Approach" in *Physical Review D* 73 (2006): 123527—as the completion of the first stage of the development of top-down cosmology. We first used the term *top-down cosmology* in a publication in S. W. Hawking and Thomas Hertog, "Why Does Inflation Start at the Top of the Hill?" *Physical Review D* 66 (2002): 123509, but this was well before we had any sort of coherent implementation of the idea.

20. Top-down cosmology echoes Dirac on this point. See note 10 in chapter 5 and, as we shall see shortly, Lemaître.

21. James B. Hartle, S. W. Hawking, and Thomas Hertog, "The No-Boundary Measure of the Universe," *Physical Review Letters* 100, no. 20 (2008): 201301.

22. Curiously, Darwin appears to have been reluctant to discuss the origin of life. In 1863, he opined in a letter to his friend Joseph Dalton Hooker that contemplating the origin of life was "mere rubbish thinking" and that "one might as well think of the origin of matter." Today, of course, we do precisely that.

23. Top-down cosmology circumvents the paradoxical loss of predictivity of the multiverse because the theory, thanks to its quantum roots, predicts relative probabilities of different wave fragments. When quantum cosmologists say that two properties of the universe are correlated, they mean that the probability is high for wave fragments in which both properties emerge through cosmological evolution. We elaborated on top-down predictions in "Local Observations in Eternal Inflation," in: James B. Harle, S. W. Hawking, and Thomas Hertog, *Physical Review Letters* 106 (2021): 141302. I remember Stephen was furious at the time that *Physical Review Letters* made us change the title of our paper. He was really fond of "Eternal Inflation without Metaphysics," the title of the manuscript we had submitted, which reflected Stephen's growing conviction that the eternally inflating multiverse would not survive a proper quantum outlook onto the universe.

24. Among the physicists who made important contributions to the further development of Everettian quantum mechanics are Robert Griffiths and Roland Omnès, and also Erich Jos, Dieter Zeh, and Wojciech Zurek.

25. Decoherent histories quantum mechanics distinguishes between fine-grained and coarse-grained histories of a system. Fine-grained histories describe all possible pathways of a system—be it a single particle, a living organism, or the universe as a whole—tracked in exquisite detail. However, that enormous level of detail also means that fine-grained histories don't decohere from one another and hence have little meaning on their own. This is where coarse-grained histories come in. Coarse-grained histories are fine-grained histories that are bundled together as it were to form a single (coarse-grained) history. Coarse-grained histories that ignore enough de-

tails of a system's evolution decohere from each other and thus have an independent existence with, for example, meaningful probabilities. But what are the fine-grained histories that should be bundled together? Said differently, what is the collection of coarse-grained histories that one should retain? This is determined by the features of the system that one wants to describe or predict. The level of coarse-graining, that is, is intimately related to the questions one asks of a system. And this is how decoherent histories quantum mechanics integrates observership in its framework.

26. Lemaître, "Primaeval Atom Hypothesis."
27. Charles W. Misner, Kip S. Thorne, and Wojciech H. Zurek, "John Wheeler, Relativity, and Quantum Information," *Physics Today* (April 2009): 40–50.

CHAPTER 7: TIME WITHOUT TIME

1. "Brane New World" started when Stephen returned to Cambridge from the U.S. sometime in the Spring of 1999. He rolled into our office to declare there was a paper to be written, and, paraphrasing Miranda in *The Tempest,* that it should be called *brane new world,* leaving us momentarily in the dark what, exactly, the paper should be about. A key question at the time was whether membrane-like universes with an invisible fourth space dimension could emerge from some kind of big bang origin. The paper "Brane New World," which we published in *Physical Review D* 62 (2000) 043501, eventually showed that in Stephen's no-boundary proposal for the origin of the universe, such (mem)brane worlds can come into being from nothing, in a process of quantum creation. Moreover, we found that the extra dimension perpendicular to the membrane, even though we can't directly observe it, could leave a subtle imprint in the cosmic microwave background fluctuations within the membrane, offering the hope that one day we might be able to test indirectly whether we live in a brane world.

2. Stephen's books often included some of his latest research indeed. His 1983 no-boundary theory was the highlight of *A Brief History of Time,* and our first top-down ideas featured in *The Grand Design* of 2010. "Brane New World" inspired the last chapter of *The Universe in a Nutshell,* where Stephen likened the birth of membrane-like universes to the creation of bubbles of steam in boiling water. This back-and-forth between his research and his popular science writings was central to his scholarly practice and reflects, I believe, his firm conviction that science, including new cutting-edge insights, should be part of our culture if it is to change the world for the better. All this meant that I wasn't the least surprised when, shortly before his death, Stephen told me it was time for a new book—this book.

3. S. W. Hawking and Thomas Hertog, "A Smooth Exit from Eternal Inflation?" *Journal of High Energy Physics* 4 (2018): 147.

4. S. W. Hawking, "Breakdown of Predictability in Gravitational Collapse," *Physical Review D* 14 (1976): 2460.

5. One might, of course, have thought that Hawking's semiclassical methods aren't suited to analyze how information escapes from an evaporating black hole. After all, black holes harbor a singularity where the semiclassical theory breaks down. However, Don Page of the University of Alberta clarified that the information puzzle isn't so much about what happens all the way at the end of a black hole's life, when the singularity surely comes into play, but about the long road toward it. Page performed his own thought experiment in which he studied the total amount of quantum entanglement between the inside of a black hole and the Hawking radiation outside. This is captured by a quantity known as the entanglement entropy, a quantum version of entropy devised by the mathematician John von Neumann that measures one's lack of information about the precise wave function of a quantum system. At the start of the evaporation process the entanglement entropy is obviously zero, since the black hole hasn't yet emitted any radiation to be entangled with. As Hawking radiation trickles out, the entanglement entropy between the hole and the radiation grows, since the emitted particles are entangled with their partner particles behind the horizon. Page reasoned that if information is to be preserved, this trend better reverse at some point so that the entanglement entropy is zero again at the end when there is no longer a black hole. He concluded that over time, the entanglement entropy should follow a curve shaped like an inverted V, with the transition point roughly halfway through the evaporation process. Since the black hole is still large at that point, Stephen's semiclassical framework should apply, for it has no business breaking down in the relatively low-curvature environment near the horizon of a large black hole. Yet there is nothing in Hawking's semiclassical calculation that can bend down the entanglement entropy curve. According to Hawking, the entanglement entropy just keeps on rising. This rising behavior sharpens the paradox. The assumption that putative quantum gravity effects get all information out just before the black hole disappears suddenly appears much less plausible. Page's refinement of Hawking's thought experiment shows that the black hole information problem is a paradox *within* the semiclassical framework of gravity. Page published his analysis "Average entropy of a subsystem" in *Physical Review Letters* 71 (1993): 1291.

6. S. W. Hawking, "Black Holes Ain't as Black as They Are Painted," The Reith Lectures, BBC, 2015.

7. Edward Witten, "Duality, Spacetime and Quantum Mechanics," *Physics Today 50*, 5, 28 (1997).

8. Maldacena arrived at his holographic duality by considering, from two different perspectives, the properties of a tightly stacked collection of three-dimensional membranes, which theorists call three-branes. Earlier, the lucid theorist Joe Polchinski had realized that such branes in M-theory are special loci that the endpoints of strings making up matter "particles" are attached to. Strings can move freely through branes but can't leave branes. The only exception to this rule are the strings responsible for gravity, because these are closed loops with no endpoints, so branes can't trap them. Physically this means that, in string theory, gravity necessarily leaks off branes, propagating into all spatial dimensions, whereas matter can be confined to branes. Now from one viewpoint, the intrinsic point of view that looks at the dynamics of strings moving through the branes, Maldacena found that the stack of three-branes is described by a quantum field theory living in three spatial dimensions—the three dimensions making up the three-branes. But next Maldacena considered the very same stack of three-branes from an exterior viewpoint, looking at how the stack as a whole impacts its environment. Looking at the stack this way, Maldacena found that it is basically a gravitating system. The branes have mass and energy and therefore they bend spacetime in their vicinity. Moreover, the curved spacetime generated by the branes turns out to extend into an additional direction orthogonal to the branes with the shape of an AdS space. Both perspectives appear radically different. However, Maldacena reasoned, since they describe one and the same physical system, they should ultimately be the same. That is, they should be dual to each other. Thus Maldacena arrived at a holographic duality relating gravity and string theory in curved AdS space to quantum field theory (QFT) on the boundary surface. Maldacena published his striking analysis in a paper "The Large N limit of superconformal field theories and supergravity" in *Advances in Theoretical and Mathematical Physics* 2 (1998), 231–52.

9. The general idea that the gravity side of the holographic duality involves a sum over interior geometries goes back to the early days of the duality. When Witten first proposed that black holes in an AdS universe have a dual description in terms of a hot bath of quarks and gluons moving about in the boundary world, he also noticed that there was a second interior geometry without a black hole floating around in his calculations. When the quark soup is hot, the no-black-hole interior was lying low. Its amplitude in the wave function was negligibly small. But when Witten lowered the temperature of the soup (as a thought experiment!), he noticed that its

composition changed, with quarks clumping together and forming tightly bound composite particles like protons or neutrons. On the gravity side, this transition from hot to cold corresponds to the no-black-hole interior coming to dominate the interior geometry with a black hole. So by changing the temperature of the particle stew on the boundary surface, either one or another geometry comes to the fore in the interior—a vivid illustration of Feynman's superposition of spacetimes. Witten published his analysis, "Anti-de Sitter space, thermal phase transition, and confinement in gauge theories" in *Advances in Theoretical and Mathematical Physics* 2 (1998), 253.

10. S. W. Hawking, "Information Loss in Black Holes," *Physical Review D* 72 (2005): 084013.

11. Geoffrey Penington, "Entanglement Wedge Reconstruction and the Information Paradox," *Journal of High-Energy Physics 09* (2020) 002; Geoff Penington, Stephen H. Shenker, Douglas Stanford, "Replica wormholes and the black hole interior," *JHEP 03* (2022) 205; Ahmed Almheiri, Netta Engelhardt, Donald Marolf, Henry Maxfield, The entropy of bulk quantum fields and the entanglement wedge of an evaporating black hole," *JHEP 12* (2019) 063.

12. Figure from John Archibald Wheeler, "Geons," *Physical Review* 97 (1955): 511–36.

13. Over the years a great many theorists have contributed to the development of a holographic duality for expanding universes such as de Sitter space, a collective effort which continues today. The first published musings on a dS–QFT duality go back to the early 2000s and include works of Andrew Strominger, and of Vijay Balasubramanian, Jan de Boer, and Djordje Minic. The universal wave function perspective on this duality was pioneered in research papers such as "Non-Gaussian features of primordial fluctuations in single field inflationary models" by Maldacena [*Journal of High-Energy Physics* 05 (2003): 013], in "Holographic No-Boundary Measure" by Hartle and Hertog [*Journal of High-Energy Physics* 05 (2012): 095], and in "Wave function of Vasiliev's universe: A few slices thereof" by Dionysios Anninos, Frederik Denef and Daniel Harlow [*Physical Review D* 88 (2013) 084049].

14. Georges Lemaître, "The Beginning of the World from the Point of View of Quantum Theory."

15. John Archibald Wheeler, "Information, Physics, Quantum: The Search for Links," in *Proceedings of the 3rd International Symposium on Foundations of Quantum Mechanics,* ed. Shun'ichi Kobayashi (Tokyo: Physical Society of Japan, 1990), 354–58.

16. The no-boundary wave goes to zero at the bottom of the bowl-like geometries that describe the origin of the universe. This was one of the defining

properties of the theory when Jim and Stephen first proposed it. Holography gives an information-theoretic interpretation of this feature.

17. S. W. Hawking, "The Origin of the Universe," in *Proceedings of the Plenary Session, 25–29 November 2016*, eds. W. Arber, J. von Braun, and M. Sánchez Sorondo, (Vatican City, 2020), Acta 24.

CHAPTER 8: AT HOME IN THE UNIVERSE

1. Arendt's essay resonates with the prologue and the latter part of her book *The Human Condition*. It was also republished, with minor edits, in the second edition of *Between Past and Future: Eight Exercises on Political Thought* (New York: Viking Press, 1968).

2. On the one hand, Arendt reasons, man is earthly, born into the world, dealing with fate and fortune and with elements beyond his control. On the other hand, man is an artificer, who can to some extent remake the world. The seeds of human freedom, in Arendt's thinking, lie in the confluence of these two competing forces.

3. Werner Heisenberg, *The Physicist's Conception of Nature*, 1st American ed. (New York: Harcourt, Brace, 1958).

4. From their writings on the subject, it is difficult to deduce whether pioneers like Dirac or Lemaître already envisaged the origin of the universe as a sort of epistemic limit too. However, shortly after this manuscript was completed the VRT, the Flemish public broadcaster, found in its archives a long-lost interview of Lemaître conducted by Jerome Verhaeghe in 1964, in which he reflects on his primeval atom hypothesis of 1931 and elaborates precisely on this point. Lemaître evokes quite clearly the idea that the "atom," as he envisions it, doesn't merely represent the beginning of time but a more profound origin that cannot be reached by thought, "an inaccessible beginning that stands just before Physics."

5. By applying computational algorithms, we can compress data and store them in a shorter message. Take, for example, the orbits of the planets. One can describe these by specifying the positions and momenta of all planets at a series of moments of times, but this message is compressible to a statement of the positions and momenta at one time, combined with Newton's equations of motion. Moreover, the data of many different gravitating systems can be compressed into a message involving the same equations of motion. This is what gives Newton's equations their universal law–like character. This is quite different, however, from endowing those equations with an independent objective existence that supersedes the cosmos.

6. In the grand scheme of quantum cosmology, the distinction between levels

of evolution isn't fundamental but arises because one zooms in on different kinds of branchings in the universal wave function. Higher levels of evolution are concerned with questions conditioned not only on the wave function but also on the specific outcomes of branching processes leading up to that level. In order to study abiogenesis on Earth four billion years ago, for example, one asks chemical questions of the universal wave function. One therefore zooms in on branchings that are concerned with that level. To do so, one must supply the outcome of the lower levels of cosmological, astrophysical, and early geological evolution in addition to a model of the wave function itself.

7. S. W. Hawking, "Gödel and the End of Physics."

8. Robbert Dijkgraaf, "Contemplating the End of Physics," *Quanta Magazine* (November 2020).

9. To support a degree of optimism, there must be particularly improbable past evolutionary steps. The rate of star formation and the abundance of exoplanets orbiting other stars indicate that, in all likelihood, the physical conditions do not present a major bottleneck. This is the biofriendly character of the physical laws again. But some of the steps associated with biological evolution remain notoriously uncertain. Evolutionary biologists have identified about seven hard trial-and-error steps that are plausible candidates for major bottlenecks on the way to everlasting life. These include abiogenesis, the formation of complex eukaryotic cells, sexual reproduction, multicellular life, and the emergence of intelligence. In the next decade or so, we stand to learn more about the likelihood of some of these transitions, from missions to Mars and observations of the atmospheres of exoplanets. If scientists find multicellular life on Mars (provided it evolved independently) or signatures of primitive life in the chemical composition of exoplanetary atmospheres, these discoveries would eliminate some of the steps in our past as candidates for improbable transitions, further sharpening Fermi's paradox.

INDEX

Page numbers of illustrations appear in italics.

ABOUT THE AUTHOR

THOMAS HERTOG is an internationally renowned cosmologist who was for many years a close collaborator of Stephen Hawking. He received his doctorate from the University of Cambridge and is currently professor of theoretical physics at the University of Leuven, where he studies the quantum nature of the big bang. He lives with his wife and their four children in Bousval, Belgium.

ABOUT THE TYPE

This book was set in Minion, a 1990 Adobe Originals typeface by Robert Slimbach. Minion is inspired by classical, old-style typefaces of the late Renaissance, a period of elegant and beautiful type designs. Created primarily for text setting, Minion combines the aesthetic and functional qualities that make text type highly readable with the versatility of digital technology.